COLLECTION MANAGEMENT

12/11	4	9/11

SPLIT-SECOND PERSUASION

SPLIT-SECOND
PERSUASION

The ANCIENT ART and NEW SCIENCE
of CHANGING MINDS

Kevin Dutton, Ph.D.

HOUGHTON MIFFLIN HARCOURT
BOSTON   NEW YORK   2011

First U. S. edition

Copyright © 2010 by Kevin Dutton

www.hmhbooks.com

First published in Great Britain in 2010 by William Heinemann

Library of Congress Cataloging-in-Publication Data
Dutton, Kevin.
Split-second persuasion : the ancient art and new science of changing
   minds / Kevin Dutton. — 1st U.S. ed.
   p. cm.
   "First published in Great Britain in 2010 by William Heinemann" — Includes
bibliographical references and index.
   ISBN 978-0-15-101279-4
   1. Persuasion (Psychology) 2. Change (Psychology) 3. Influence
(Psychology) I. Title.
   BF637.P4D88 2011   153.852
   153.8'52 — dc22   2010005739

Book design by Melissa Lotfy

Printed in the United States of America

DOC 10 9 8 7 6 5 4 3 2 1

Illustration Credits appear on page 283.

Contents

Author's Note vi

Introduction 1

1. THE PERSUASION INSTINCT 12

2. FETAL ATTRACTION 32

3. MIND THEFT AUTO 65

4. PERSUASION GRANDMASTERS 99

5. PERSUASION BY NUMBERS 128

6. SPLIT-SECOND PERSUASION 157

7. THE PSYCHOPATH — NATURAL BORN PERSUADER 195

8. HORIZONS OF INFLUENCE 226

Appendix 1: Key Stimuli and Stereotyping: Socioeconomic Status 257

Appendix 2: Asch's Supplementary Traits 259

Notes 261

Acknowledgments 281

Illustration Credits 283

Index 285

Author's Note

For legal (and sometimes personal) reasons, the names and identifying details of certain people featured in this book (Keith Barrett, Shaffiq Khan, Pat Reynolds, Vic Sloan, Greg Morant, and my friend Paul) have been changed. All of those in question granted me permission to recount their ingenious exploits on the understanding that "nothing came back to them." In the case of one of these, con man Keith Barrett, attributes from several real-life individuals were combined in order to avoid cramming an inordinate number of these colorful characters into ninety thousand words. Nothing was exaggerated, and all factual details are based on the author's firsthand knowledge and empirical research.

The author is also delighted to take sole responsibility for the grammatical turbulence you will occasionally run into in this book — overuse of dashes, split infinitives. And beginning sentences with "and."

SPLIT-SECOND PERSUASION

Introduction

One evening, at the close of a lavish state banquet for Commonwealth dignitaries in London, Winston Churchill spots a fellow guest about to steal a priceless silver saltcellar from the table. The gentleman in question slips the precious artifact inside his dinner jacket, then quietly makes for the door.

What is Churchill to do?

Caught between loyalty to his host and an equal and opposite desire to avoid an undignified contretemps, he suddenly has an idea. With no time to lose, he quickly picks up the matching silver pepper-pot and slips it inside his own jacket pocket. Then, approaching his "partner in crime," he reluctantly produces the condimentary contraband and sets it down in front of him.

"I think they've seen us," he whispers. "We'd better put them back."

AIR HOSTESS: Please fasten your seatbelt before takeoff.
MUHAMMAD ALI: I'm Superman. Superman don't need no seatbelt!
AIR HOSTESS: Superman don't need no airplane!

WILD HORSES

It's six o'clock on a dead December evening in North London. Two men stand drinking in a bar in Camden Town. They finish their pints, set them back down on the counter, and look at one another. Same again? Sure, why not? Though they do not know it yet, these two men are about to be late for a dinner engagement. In an Indian restaurant far across town, another man sits waiting for them. Some casual, low-grade Parkinson's loosens the wires in his flickering right hand, and he's tired. He is wearing a brightly colored new tie which he has bought specially for the occasion, and which took him half an hour to do up. It has teddy bears on it.

It is Sunday. The man in the restaurant watches the rain gust darkly against the low-lit windows. Today is his son's birthday. In the bar in Camden Town the other men watch, too, as the rain strobes amber in the desolate glow of the street lamps, glazing the liquid pavements with a slick of neon gold. Time to head off, they say. To the train. To the restaurant. To the man who is sitting there waiting. And so they leave.

They arrive late, by almost three-quarters of an hour. Somehow, they find this amusing. They have misjudged, in hindsight by quite some considerable margin, the length of time required to consume four pints of beer and then negotiate the outer reaches of the Piccadilly and Northern lines. Instead of setting aside a couple of hours for the venture, they have allowed something in the region of ten minutes. To make matters worse, they are drunk. On their arrival at the restaurant, things do not go well.

"Late *again?*" the man who has been waiting for them inquires sarcastically. "You'll never learn, will you?"

The response is as vehement as it is instant — a million age-old grievances all rolled into a single defining moment. One of the newcomers, the smaller of the two by quite some way, turns around and walks straight back out of the restaurant. It's the son. But not before he has uttered a few well-chosen words of his own.

And so there he is, the little man. A couple of minutes earlier, rattling down a tube line heading west, he had been looking forward to a simple birthday dinner with his father and his best friend. Now he is alone under derelict December skies, hurtling back along the pavement in the direction of the tube station. Freezing cold and soaking wet because he's forgotten to pick up his coat. Funny how quickly things can change.

When the little man arrives at the station, he is seething. He stands for a few moments at the ticket barrier trying to locate his pass, and thinks to himself that wild horses wouldn't be able to drag him back to that restaurant. The station concourse is flooded and there is no one around. But then he hears something coming from the street: the sound of approaching footsteps. Suddenly, out of nowhere, there's the big man. Having legged it from the restaurant to the station, he slumps against a pillar by the entrance. The little man moves away.

"Wait!" says the big man, when he's finally got his breath back.

The little man isn't interested.

"Don't even think about it," he says, raising his hand an inch, maybe two inches above his head. "I've had it up to here with his snide remarks!"

"But wait!" says the big man again.

The little man is getting angrier by the second.

"Look," he says, "you're wasting your time. Just go back to him. Go back to the restaurant. Go wherever you like. Just get out of my face!"

The big man is worried that the little man is going to hit him.

"OK," he says. "OK. But before I go, will you just let me say one thing?"

Silence. The rain turns crimson as some traffic lights change at the crossing by the station entrance.

Just to get rid of him, the little man relents.

"Go on then," he says. "What is it?"

There's a moment of truth as the two of them look at each other — the big man and the little man — across the barrier. The little man notices that a couple of buttons have fallen off the big man's overcoat, and that his woolen bobblehat is lying on the ground in a puddle some distance away. Must've been quite a dash, the little man thinks to himself. From the restaurant to the station. And then he remembers something. Something the big man had told him once. About how his mother had knitted him that hat, one Christmas.

The big man holds out his arms — a gesture of helplessness, openness, perhaps both.

And then he says it.

"When was the last time you ever saw me run?"

The little man opens his mouth, but finds himself treading words. Suddenly, he's in trouble. The problem is that the big man weighs close to 28 stone. Though they've been friends for quite some time, the little man has *never* seen the big man run. Which is actually kind of funny. In fact, by his own admission, the big man has trouble walking.

The more the little man thinks about it, the more he finds himself struggling for an answer. And the more he struggles, the more he feels his anger ebb away.

Eventually, he says: "Well, never."

There's a period of silence. Then the big man puts out a hand.

"Come on then," he says. "Let's go back."

And so they do.

When they get back to the restaurant, the little man and his father both say sorry to each other and the three wiser, if not entirely wise, men sit down to have dinner together. For the second time. Nobody's talking miracles, but they sure as hell are thinking it. The big man had lost some buttons. And the woolen hat that his mother had made him would never be the same again. But somehow, somewhere, in the wind and the rain and the cold, he had traded them for something better.

There was nothing, the little man thinks to himself, that anyone could have said to him in that tube station that would have made him go back to the restaurant. Wild horses could not have dragged him. Yet the big man had done exactly that with just ten simple words. Words that had come from a kingdom south of consciousness:

"When was the last time you ever saw me run?"

Somehow, in the depths of that London winter, the big man had drummed up some sunshine.

HONESTY'S THE BEST POLICY

Here's a question for you. How many times a day do you think someone tries to persuade you? What to do. What to buy. Where to go. How to get there. And I'm talking from the moment you wake up in the morning to the moment your head hits the pillow again in the evening. Twenty? Thirty? That's what most people say when asked this question — so try not to feel too bad about what's coming next. In actual fact — get ready — estimates tend to hover around the four hundred mark! Comes as a bit of a shock at first, doesn't it? But let's think about it for a minute. Go through the options. What molecules of influence can infiltrate the pathways of our brain?

Well, for a start there's the advertising industry. TV. Radio. Billboards. Web. How many times a day do you think you see an ad? Right — quite a few. Then there's all the other stuff we see. The man selling hot dogs on the street corner. The policeman directing traffic. The religious guy with the sandwich board in the middle of that traffic. And then, of course, there's the little guy in our heads who's almost always banging on about something. OK, we might not actually see him but we certainly hear him often enough. Starts to mount up when you think about it, doesn't it? And believe me — we haven't even started yet.

When it comes down to it, we take all this for granted, don't we? Which is why, when asked how many times a day people try to get us to do stuff, we say twenty or thirty instead of four hundred. But there's an even more fundamental question here, one that's seldom even considered.

Where does such persuasion come from — I mean, originally? A lot is written about the origins of mind, but what about the origins of *changing* minds?

Let's imagine an alternative society to the one just described — a society in which coercion, not persuasion, is the primary tool of influence. Just think what it would be like if every time we decided *not* to buy a hot dog, the vendor on the street corner came charging after us with a baseball bat. Or if, when we shot past the speed gun at 80 mph, some death-dealing sensor riddled our windscreen with bullets. Or if when we didn't sign up with the "right" political party, or the "right" religion — have the "right" color skin even — we suffered the consequences later.

Some of these scenarios are, I would guess, easier to imagine than others. But the point I'm making here is simple. It's largely because of persuasion that we have a "society" at all. There have been various attempts at various times to challenge such a notion. But each, at some stage, has fallen decidedly short. Persuasion is what keeps us alive. Often, quite literally.

Take the following instance. In the autumn of 2003, I fly to San Francisco for a conference. Pressed for time before leaving Cambridge, I decide, because I am insane, against the time-honored wisdom of booking a hotel in advance, and opt instead to seek one out when I get there: a cheap, if somewhat frenetic, establishment in a neighborhood so dangerous that even the serial killers go round in pairs.

Every morning on leaving my brothel — I mean, hotel — and every evening upon my return, I run into the same bunch of guys huddled around the newspaper stand outside: a Vietnam veteran with six months to live, a Brazilian hooker more down on her luck than anything more lucrative, and a flotilla of hungry and homeless who between them have taken more hits than a Paris Hilton slumber party on YouTube. All have had their fair share of misfortune. All their misadventures. And all stand despondently on the pavement, their windswept placards and rain-sodden signboards stacked up dejectedly beside them.

Now I'm not saying that these guys didn't need the money. They did. But after a week of small talk and slowly getting acquainted, it had reached the point where our fortunes had all but reversed—and *I* was asking *them* for cash. Most of the posse I was on first-name terms with, and after shelling out generously for the first couple of days any desire to swell their coffers further had disappeared faster than a Bernie Madoff hedge fund.

Or so I thought.

Then one night, toward the end of my stay, I notice a guy I haven't seen before. By this stage I'm building up a bit of immunity to the hard luck stories, and as I pass I give no more than a fleeting glance at the dog-eared piece of cardboard he's holding out in front of him. Yet no sooner does the message catch my eye than I'm poking around inside my coat pocket, looking for something to give him. And not loose change but something more substantial. A mere five words has got me reaching for my wallet without a moment's hesitation:

WHY LIE? I WANT BEER!

I felt I'd been legally mugged.

Back in the safety—well, *comparative* safety—of my hotel room I sit there thinking about that slogan. Even Jesus would have applauded. I wasn't usually in the habit of doling out money to pissheads. Especially when, just a few feet away, more deserving causes beckoned. Yet that was precisely what I'd done. What was it about those five words that had had such an effect on me? I wondered. The guy couldn't have got the money out of me any faster had he pulled a gun from his jacket. What was it that had so cleanly, so comprehensively, yet so covertly disabled all those cognitive security systems I'd been so painstakingly intent on installing since my arrival?

I smile.

Suddenly, I'm reminded of a similar occasion many years before when I'd argued with my father in a restaurant. And then stormed off. There was no way, I'd thought to myself at the time, that I would be going back to that restaurant that night. *Wild horses could not have dragged me.* Yet a mere thirty seconds and ten words later a friend of mine had dramatically changed my mind.

There was, I began to realize, something inherent in both these incidents that was timeless, weightless, and fundamentally different from normal modes of communication. They had a transforming, transcendental, almost otherworldly quality about them.

But what exactly was it?

A SUPERSTRAIN OF PERSUASION

As a psychologist, up there in my hotel room, I felt I should have an answer to that question. But the more I thought about it, the more I struggled to come up with one. This was a question about persuasion. About attitude change. About social influence. Regular banter in the social psychology locker room — and yet, it now seemed, there was a big black hole in the literature. I was baffled. How could a total stranger clean out my wallet with just five simple words? And how, with just ten, could my best friend clean out my brain?

Usually, it works like this. If, like my friend, we want to calm someone down, or like the beggar extract money from them, we tend to take our time over it. We carefully set up our pitch. And with good reason. Minds — just ask any used car salesman — don't change easily. Nine times out of ten, persuasion is contingent on a complex combination of factors, relating not just to what we say but also how we say it. Not to mention, once said, how it's interpreted. In the vast majority of cases, influence is wrought by talking. By a nervy cocktail of compromise, enterprise, and negotiation. By wrapping up whatever it is that we want in an intricate parcel of words. But with my friend and the homeless guy it was different. With them, it wasn't so much the wrapping that did the trick as — well — the lack of it. It was the immaculate incisiveness of the influence; the raw and chastening elegance; the deft, swift touch of psychological genius that, more than anything else, gave it its power.

Or was it?

No sooner had I extricated myself from San Francisco and returned to the equally chaotic, if somewhat less predictable, milieu of Cambridge academic life than it began to dawn on me just how wide ranging a question this was. Did there exist, buried deep within the geology of rhetoric, an elixir of influence — a secret art of split-second persuasion that each of us might learn? To close that deal? To get that guy? To tip those scales just that little bit more in our favor?

Much of what we now know about the brain — the relationship between function and structure — has come about not from the study of the conventional but of the extraordinary. From extremes of behavior at odds with the everyday. Might the same also be true of persuasion? Take the Sirens in Homer's *Odyssey*: beautiful maidens whose song is so bewitching that mariners, even on pain of death, are irresistibly drawn to it. Or Cupid and his arrows. Or the "secret chord that pleased the Lord" that David plays in the Leonard Cohen song "Hallelujah." Outside the realms of mythology, might such a chord actually exist?

As my research progressed the answer to this question soon became clear. Slowly but surely, as the list of examples grew longer and the cold, digital voodoo of the statistics began to unfold, I started to piece together the elements of a brand new kind of influence. To plot the genome of a mysterious, previously unidentified, superstrain of persuasion. Most of us have some idea of how to persuade. But it's largely trial and error. We get it wrong as many times as we get it right. Yet some people, it began to emerge, really *could* get it right. With absolute precision. And not just around the coffee table. Or out on the town with friends. But in knife-edge confrontations where both stakes and emotions ran high. So who were these black belts of influence? And what made them tick? More importantly — what, if anything, could they teach the rest of us?

Here's another example. Imagine it had been you on the plane. What do you think *you* would have said at the time?

I'm on a flight (business class, thanks to a film company) to New York. The guy across the aisle from me has a problem with his food. After several minutes of prodding it around his plate, he summons the chief steward.

"This food," he enunciates, "sucks."

The chief steward nods and is very understanding. "Oh my God!" he schmoozes. "It's such a pity . . . You'll never fly with us again? How will we ever get the chance to make it up to you?"

You get the picture.

But then comes something that totally changes the game. That doesn't just turn the tables, it kicks them over.

"Look," continues the man (who was, one suspected, quite used to continuing), "I know it's not your fault. But it just isn't good enough. And you know what? I'm so fed up with people being *nice*!"

"IS THAT RIGHT, YOU FUCKING DICK? THEN WHY THE FUCK DON'T YOU FUCKING SHUT UP, YOU ARSEHOLE?"

Instantly, the whole cabin falls silent (at which point, in an amusing coincidence, the "Fasten Seat Belt" sign also comes on). Who the hell was *that*?

A guy in one of the front seats (a famous musician) turns round. He looks at the guy who's complaining, then winks at him.

"Is that any better?" he inquires. "'Cos if it ain't, I can keep going . . ."

For a moment, nobody says anything. Everyone freezes. But then, as if some secret neural tripwire had suddenly just been pulled, our disgruntled diner . . . smiles. And then he laughs. And then he *really* laughs. This, in turn, sets the chief steward off. And that, of course, gets us all started.

Problem solved with just a few simple words. And definitive proof, if ever any were needed, of what my old English teacher Mr. Johnson used to say: "You can be as rude as you like so long as you're polite about it."

But back to my original question. How do you think you would have reacted under such circumstances? How would you have got on? Me? Not too well, as it turned out. But the more I thought about it, the more I came to realize precisely what it was about situations like these that made them so special. It wasn't just the psychological bull's-eyes — spectacular though some of them might be. No, it was more than that. It was the individuals who scored them.

I mean, think about it. Forget about the musician for a moment. In the absence of screwballs like him, air stewards (not to mention, in more hostile scenarios, policemen, members of the armed forces, professional negotiators, health care workers, and Samaritans) face such dilemmas as these every day of their lives. These are people who are trained in the art of persuasion; who use tried-and-tested techniques to maintain the status quo. Such techniques involve building a relationship with the other person and engaging him or her in dialogue while at the same time projecting a calm, patient, and empathic interpersonal style. Techniques, in other words, that are underpinned by social process.

But there are, quite clearly, some of us who are simply "naturals." Who do not need to train. Who are, in fact, so good, so extraordinarily different, that they have a gift for turning people around. Not through negotiation. Or dialogue. Or the rules of give-and-take. But with just a few simple words.

Sound crazy? I know. Back when the idea first came to me, I thought the same. But not for long. Soon I began to unearth a tantalizing body of evidence — circumstantial, anecdotal, allusive — which suggested the possibility that there really might be black belts in our midst. And, what's more, they might not all be good guys.

CRACKING THE CODE OF PERSUASION

This, then, is a book about persuasion. But it's a book about a special kind of persuasion — *split-second persuasion* — with an incubation period of sec-

onds and an evolutionary history just a little bit longer. Incongruity (or surprise) is obviously a key component. But that's just the beginning. Whether
we take what's on offer or leave it on the table depends on four additional
factors: simplicity, perceived self-interest,* confidence, and empathy — factors as integral to persuasion in the plant and animal kingdoms as they
are to the scams of some of the world's most brilliant con artists. Together,
this five-part cocktail of influence — SPICE — is lethal. And all the more so
when taken straight: undiluted by rhetoric, uncontaminated by argument.

Winston Churchill certainly knew as much. And as for the air hostess who once took on The Greatest — I doubt Muhammad Ali ever took a
cleaner shot in his life.

It's a kind of persuasion that can get you whatever you want. Reservations. Contracts. Bargains. Babies. *Anything.* In the *right* hands. But which in
the wrong hands can prove disastrous. As brutal and deadly as any weapon
that's out there.

The journey begins with a simple idea: that some of us are better at the
art of persuasion than others. And that with persuasion, just as with everything else, there exists a spectrum of talent along which each of us has our
place. At one end are those who always "put their foot in it." Who seem not
only to get the wrong end of the stick, but sometimes the wrong *stick* even.
At the other we have the split-second persuaders. Those who exhibit an uncanny, almost preternatural, propensity for "getting it right."

In the pages that follow, we plot the coordinates of this mysterious
strain of persuasion. Slowly but surely, as we cast the net of empirical enquiry farther and farther afield, beyond the familiar reefs of social influence
to the deeper, less-charted waters of neonatal development, cognitive neuroscience, mathematics, and psychopathology, we navigate theories about
the chimeric art of persuasion that slowly begin to converge. That gradually distil into a single, definitive formula. Our journey uncovers a treasure
trove of questions:

- What do newborn babies and psychopaths have in common?
- Has our ability to change minds, like the mind itself, evolved?

---

*This refers to the self-interest of the target. Persuasion, of course, is not always in the
target's *actual* interest. But if the target perceives it to be so, then the attempt is far more
effective.

- What secrets do the all-time greats of persuasion and the grandmasters of martial arts have in common?
- Is there a "persuasion pathway" in the brain?

The answers will amaze you. And will definitely, next time you go for that upgrade, help you get it.

1

The Persuasion Instinct

JUDGE: I find you guilty as charged and hereby sentence you to seventy-two hours' community service and a fine of £150. You have a choice. You can either pay the full amount within the allotted three-week period or pay £50 less if you settle immediately. Which is it to be?

PICKPOCKET: I only have £56 on me at the moment, Your Honor. But if you allow me a few moments with the jury, I'd prefer to pay now.

A policeman on traffic duty pulls a motorist over for speeding.

"Give me one good reason why I shouldn't write you a ticket," he says.

"Well," says the driver, "last week my wife ran off with one of you guys. And when I saw your car, I thought you were bringing her back."

A SPEWRIOUS TALE?*

In 1938, in Selma, Alabama, a physician by the name of Drayton Doherty was summoned to the bedside of a man called Vance Vanders. Six months earlier, in a graveyard in the dead of night, Vanders had bumped into a witch-doctor and the spook had put a curse on him. A week or so later Vanders got

---

*Actually, not. The case is documented in Clifton K. Meador's *Symptoms of Unknown Origin: A Medical Odyssey* (Nashville: Vanderbilt University Press, 2005). It is also recounted in an article by Helen Pilcher, "The Science and Art of Voodoo: When Mind Attacks Body," *New Scientist*, May 13, 2009, 30–33.

a pain in his stomach, and decided to take to his bed. Much to the distress of his family, he'd remained there ever since.

Doherty gave Vanders a thorough examination, and grimly shook his head. It's a mystery, he said, and shut the door behind him. But the next day he was back.

"I tracked down the witchdoctor and lured him back to the graveyard," he announced. "When he arrived I jumped on him, pinned him to the ground, and swore that if he didn't tell me the exact nature of the curse he'd put on you, and give me the antidote, I would kill him on the spot."

Vanders's eyes widened.

"What did he do?" he asked.

"Eventually, after quite a struggle, he relented," Doherty continued. "And I must confess that, in all my years in medicine, I've never heard anything like it. What he did was this. He implanted a lizard egg inside your stomach — and then caused it to hatch. And the pain you've been feeling for the last six months is the lizard — it's been eating you alive!"

Vanders's eyes almost popped out of his head.

"Is there anything you can do for me, Doctor?" he pleaded.

Doherty smiled reassuringly.

"Luckily for you," he said, "the body is remarkably resilient and most of the damage has been largely superficial. So we'll administer the antidote the witchdoctor kindly gave us, and wait and see what happens."

Vanders nodded enthusiastically.

Ten minutes later, his patient vomiting uncontrollably from the powerful emetic he'd given him, Doherty opened his bag. Inside was a lizard he'd bought from the local pet shop.

"Aha!" he announced with a flourish, brandishing it by the tail. "*Here's* the culprit!"

Vanders looked up, then retched violently again. Doherty collected his things.

"Not to worry," he said. "You're over the worst of it and will soon pick up after this."

Then he left.

Sure enough, for the first time in ages, Vanders slept soundly that night. And when he awoke in the morning he had eggs and grits for breakfast.

Persuasion. No sooner is the word out than images of secondhand car dealers, mealy-mouthed politicians, schmoozers, cruisers, and a barrel-load of life's other users and abusers come padding—brothel-creepers and smoking jackets at the ready—across the dubious neuronal shag piles of our minds. It's that kind of word. Though undoubtedly one of social psychology's hippest, most sought after neighborhoods, persuasion also has a dodgy, downbeat reputation: an area of Portakabins and bars, sleazy garage forecourts, and teeming neon strips.

Which, of course, is where you often find it at work.

But there's more to persuasion than just cheap talk and loud suits. Or, for that matter, loud talk and cheap suits. A witchdoctor and physician go head to head (quite literally) over the health of a local man. The witchdoctor deals what appears to be a knockout blow. His opponent rides in and effortlessly turns the tables. This extraordinary tale of a shaman and a split-second persuader encapsulates influence in its simplest, purest form: a battle for neural supremacy. Yet where does persuasion come from? Why does it work? Why is it possible that what is in my mind, when converted into words, is able to change what's in yours?

The ancient Greeks, who seemed to have a god for more or less everything, had one, inevitably, for persuasion. Peitho (in Roman mythology, Suadela) was a companion of Aphrodite and is often depicted in Greco-Roman culture with a ball of silver twine. These days, of course, with Darwin, game theory, and advances in neuroimaging, we see things a little differently. And with the gods up against it and the Greeks more interested in basketball, we tend to look elsewhere for affirmation. To science, for instance. Or Oprah.

In this chapter we turn our attention toward evolutionary biology—and discover that persuasion has a longer family history than either we, or the gods, might have realized. We go in search of the earliest forms of persuasion—prelinguistic, preconscious, prehuman—and arrive at a startling conclusion. Not only is persuasion *endemic* to earthly existence, it's also *systemic,* too; as much a part of the rhythm of the natural order as the emergence of life itself.

PURRSUASION

Note to architects who are currently in the process of designing modern, shiny, glassy buildings for affluent, leafy, tree-lined neighborhoods: spare a thought for the local bird population.

In 2005, the MRC Cognition and Brain Sciences Unit in Cambridge was having trouble with kamikaze pigeons. The courtyard of a brand new extension block was proving a black spot for avian suicides, with as many as ten birds a day dive-bombing the window of the state-of-the-art lecture theater. It didn't take long to fathom out the reason. Reflected in the glass were the surrounding trees and bushes. And the birds — like some architects I could mention — couldn't tell the difference between appearance and reality. What to do?

In contrast to the diagnosis, the remedy proved elusive. Curtains, pictures — even a scarecrow — all came to nothing. Then one day, Bundy Mackintosh, one of the researchers who worked in the building, had an idea. Why not talk to the birds in their own language?

So she did.

Mackintosh cut, out of a sheet of colored cardboard, the profile of an eagle and then stuck it in the window. Deep in their brains, she reasoned, the birds would have a console — a sort of primitive mental dashboard on which, silhouetted as birds of prey, would appear a series of hazard warning lights. As soon as one of these predators came into view, the corresponding light would immediately flash up red — and an ancient evolutionary forcefield would suddenly engulf the unit, repelling the birds and diverting them from the danger.

Problem solved.

Talking to animals in their own native tongue (as Bundy Mackintosh did in a very simple way with her cardboard and scissors) involves empathy. And learning the syntax of biological vernacular. And if you think it's just humans who can do it, think again. Biologist Karen McComb of the University of Sussex has discovered something interesting about cats: they employ a special "solicitation purr" which hot-wires their owners to fill up their food bowls at dinnertime.

McComb and her coworkers compared cat owners' responses to different kinds of purr — and found that purrs recorded when cats were seeking food were more aversive and harder to ignore than other purrs played at the same volume. The difference is one of pitch. When cats are soliciting food, they give off a classic "mixed message" — embedding an urgent, high-pitched cry within a contented, low-pitched purr. This, according to McComb, not only safeguards against instant ejection from the bedroom (high pitch on its own) but also taps into ancient, mammalian nurturing instincts for vulnerable, dependent offspring (more on that later).

"The embedding of a cry within a call that we normally associate with contentment is quite a subtle means of eliciting a response," explains Mc-Comb, "and solicitation purring is probably more acceptable to humans than overt meowing."

Or, to put it another way, cats, without the linguistic baggage of forty thousand words (the estimated vocabulary of the average English-speaking adult), have learned a faster, leaner, more efficient means of persuading us to do their bidding — exactly the same strategy that Bundy Mackintosh hit upon to "talk" to the pigeons of Cambridge. The deployment of what is known in ethology as the *key stimulus*.

MORE THAN WORDS CAN SAY

A key stimulus is influence in its purest form. It is neat, 200 proof mind control — undiluted by language and the thought fields of consciousness — which is knocked back straight, down the hatch, like a shot. Key stimuli are simple, unambiguous, and easily understood: persuasion as originally intended. Officially, of course, the definition is somewhat different: a key stimulus is an environmental trigger that initiates, solely by its presence, something known as a fixed action pattern — a unit of innate behavior that continues, once initiated, uninterrupted to completion. But it amounts, more or less, to roughly the same thing.

Numerous incidences of key stimuli are found within the natural world, not least when it comes to mating. Some are visual, like Bundy Mackintosh's eagle silhouette. Some acoustic, like the solicitation purr. And some kinetic, like the way honeybees dance to communicate the location of a food source. Some combine all three. *Chiroxiphia pareola* is renowned for its cobalt coat, its sweet and melodious warble, and its elaborate courtship ritual (which, uniquely, involves a dominant male supported by a five-strong backing band). No, *Chiroxiphia pareola* is not the Latin for Barry White but a tropical songbird found deep in the Amazon jungle. It has a brain the size of a pea.

Chiroxiphia pareola is no member of the Seduction Community.* Yet there's nothing you can tell it about pulling. When the male of the species encounters a suitable mate he doesn't, all of a sudden, start beating around his bush. Quite the opposite, in fact. He dances straight out of it. And scores.

---

*The Seduction Community is a group of male pickup artists who employ the principles of evolutionary social psychology to attract women. The community, and its practices, are documented in Neil Strauss's *The Game: Penetrating the Secret Society of Pickup Artists* (Edinburgh: Canongate, 2005).

In certain species of frog it is sound, primarily, that makes up the language of love. The Green Treefrog is one of Louisiana's most instantly recognizable critters — especially if you're tired and trying to catch forty winks. More commonly known as the Bell Frog (because of the distinctive sound of its mating call, which resembles a ringing bell: quonk, quonk, quonk), it's equally at home in a variety of different environments such as ponds, roadside ditches, rivers, and swamps. Not to mention well-lit verandas where it feeds, among other things, on sleep deprivation.

The acoustical arsenal of the Bell Frog is actually more complicated than it appears. When calling in unison, for instance, individuals frequently coordinate their efforts — and the resulting cacophony will often emerge as a harmonious (though exasperating!) "quonkquack, quonkquack" refrain. Research has also shown that males tend to vary their calls depending on the circumstances. At dusk, for example, as a precursor to hitting the breeding pool, they will issue a preliminary "territorial" call (one that tells other males to back off), and then, while en route to the pool, resort to a rather more prickly sounding chunter as they gruffly, and somewhat slothfully, bump into each other. It's only on reaching the breeding pool that they really open up — cranking up the chorus to its anthemic "quonk quonk" finale. So resonant, in fact, is this eponymous mating call that female Bell Frogs can actually make it out from up to 300 meters away. A statistic, oddly enough, not lost on local residents.

CROAK AND DAGGER

Up until now, the influence that we've been looking at in birds and frogs has been the kind of honest, straight-down-the-line persuasion we see repeated a million times over in human society — the only difference being that these guys do it better. From finding a partner to nailing that crucial deal, success depends on speaking a common language. And they don't come any more common than the key stimulus.

But the importance of this common language in persuasion — this mutual understanding, or empathy* — is brought into even sharper focus when we consider a completely different kind of influence, mimicry: when

---

*In this context I use the term *empathy* rather loosely to refer, in the absence of consciousness, to the capacity to "connect" — to frame a communication in such a way as to maximize salience to its recipient.

a member of one species assumes or manipulates the characteristics of another (though this can also occur *intra*species) for the purposes of personal
advancement.

Let's stay with Bell Frogs for the moment. For most frogs, the dating
game is set in stone. I mean, face it — when all you can do is croak, there
isn't much room for maneuver. What tends to happen is this. The males just
sit there and croak . . . and the females, if they get lucky, come hopping. It
couldn't be any simpler. But Bell Frogs have figured something out. These
little buggers have incorporated an element of skullduggery into the proceedings, and it's by no means unusual for a deeply resonant baritone in
full quonk to be stalked, completely unawares, by a silent, shadowy cadre of
male hangers-on.

This bears testament to the steely ingenuity of natural selection. Think
about it. A hard night's quonking uses up vital energy stores. And because of
this, one of two things can happen. On the one hand, the caller might draw
a blank, and — exhausted — hail a taxi. On the other, he might get lucky
and finish up down by the breeding pool. On whichever note the evening
finally ends doesn't really matter. Observe, in either case, what happens to
the original calling site once its former occupant slopes off. Suddenly it goes
on the market. And turns, in the process, into prime location real estate for
any one of the nonquonking identity thieves to clean up in. Any unsuspecting female who shows up after the quonker has left discovers — as if nothing has changed — a nonquonking impostor in his place. But how is she to
tell the difference? Bottom line is: she can't.*

SELF BELEAF

As a weapon of persuasion mimicry is ingenious. If the key stimulus is
influence taken straight, then mimicry, you could say, is empathy taken
straight. Just like the key stimulus, there are several different kinds — not all
of which, as we've just seen with the Bell Frogs, are benign.

For a start, there's the most obvious form — visual mimicry — which is

---

*And identity theft isn't the only kind of racketeering that these double-dealing lotharios
have a hand in. Bell Frog psychopaths — of the nonquonking fraternity — routinely mug
their exhausted quonker buddies by leaping out of the shadows at the very last minute
and accosting their females: the selfsame females that their worn-out counterparts have
just spent the whole of the night serenading.

sort of what the nonquonking love rats get up to down in Louisiana. But depending on the scale of the biological forgery, and how sophisticated it is, there are also more subtle varieties that incorporate, alongside visual cues, both auditory and olfactory ones too.

A good example of this hybrid mimicry is found in plants (when I said that persuasion was integral to the natural order, I meant *all* of it). The discomycete fungus *Monilinia vaccinii-corymbosi* is a plant pathogen that infects the leaves of blueberries, causing them to secrete sweet, sugary substances such as glucose and fructose. When this happens something rather interesting occurs. With the leaves, in effect, now producing nectar — thus fraudulently impersonating flowers — they begin, like flowers, to attract pollinators, even though they actually look nothing like flowers and still, in every other respect apart from smell, resemble leaves. Natural selection then takes care of the rest. A bee drops by believing the sugar to be nectar. It slurps some up (during which time the fungus attaches itself to its abdomen) then subsequently moves on to the blueberry flower proper where it transfers the fungus to the ovaries. There, on the ovaries, the fungus reproduces — spawning mummified, inedible berries, which hibernate over the winter before going on to infect a fresh crop of plants in the spring. Clever, huh?

But the hustle doesn't end there. There is, it turns out, a whole other level to this seedy little love triangle. The olfactory emissions from the surface of the blueberry plant's leaves aren't the only ones. The infected leaves, upon analysis, also reflect ultraviolet (which, under normal circumstances, they absorb) — but which the flowers emit as a low-level come-on to insects. Suddenly, it turns out, the leaves have snitched not just one aspect of the blueberry flower's identity but two, visual and olfactory. Now that, for a common or garden fungus, really is clever.

WEB OF DECEIT

As an example of natural mimicry the discomycete fungus's antics are actually somewhat unusual. Ordinarily, rather than implicating a third party in the scam — in this case, leaves — the mimic does its own dirty work. Pygmy Owls, for instance, have "false eyes" on the reverse side of their heads, to fool predators into thinking they can, quite literally, see out of the back of them. Conversely, Owl Butterflies have eyespots on the underside of their wings so that, on suddenly flipping over, they resemble the face of an owl (see Figure 1.1). Hairstreak Butterflies go one better and, like a number of species of insect, possess filamentous "tails" at the ends of their wings. These tails, when combined with other elements of wing patination, create the distinct impression of a false head — which bamboozles predators and misdirects attack. Two heads, as they say, are often better than one.

Figure 1.1 — The distinctive, and intricately crafted, eyespots on the wings of an Owl Butterfly.

Less benign uses of distraction are glimpsed in the world of arachnids. The Golden Orb Weaver (a spider quite common in the New World) gets its colorful appellation from its dazzling golden web, which it spins (not, at first sight, the most brilliant idea in the world for rustling up dinner if you're a spider) in conspicuous, brightly lit areas.

But there's method in the Golden Weaver's madness. Research reveals that bees, contrary to common sense, find it easier to steer clear of the web when they should actually find it more difficult: when the light is poor, when the filaments are harder to see, and when the yellow pigmentation is indistinct. Why? Well, think about it. When it comes to nectar-producing flowers, which do you suppose is the most common presenting shade?

Support for such a theory comes from experiments that have ingeniously varied the color of spiders' webs. While the bees have little trouble in associating other pigments with danger — red, blue, and green, for example — and subsequently learning to avoid them, it's yellow, time and again, that poses them the greatest difficulty.

Similar zoological scams are also found in the insect world. The "honey trap" may well have been the stock-in-trade of some of Hollywood's best-known secret agents over the years, but ever wondered who thought of it first? You need look no further than the firefly. Studies have shown that female fireflies of the genus *Photuris* emit precisely the same light signals as females of the *Photinus* genus issue for mating calls. But that's not all. Research has also revealed something else. Male *Photinus* fireflies attempting to make out with these masquerading femmes fatales get quite a lot more than they bargain for. They get eaten. I had a date like that once.

IT ALL ADS UP

So far in this chapter we've been looking at how animals — and plants — "do" persuasion. How, in the absence of language, interests are served and influence is wielded. And it is, without a doubt, influence — exactly the same kind of influence that we see at work in humans. Only faster, less messy, more concentrated. How else would you describe it? Contrary to outward appearance, the Golden Orb Weaver spider doesn't have a diploma in fine art; nor does it attend night classes in interior design. And yet its web is yellow. Why? For one reason, and one reason only. To manipulate bees into doing something silly. Into doing something that they otherwise, as bees, wouldn't dream of doing. Dropping in for a visit.

It's the same with our discomycete fungus. This unscrupulous, psychopathic fungus with its dodgy botanical morals knows only too well that bees and other pollinating insects will not, in the normal run of things, touch it with a barge pole. So what does it do? It does what any other unconscionable, upwardly mobile social predator would do: enlists the help of an innocent third party and ruthlessly exploits it as a go-between. Just because there's no language involved doesn't mean to say that there's no persuasion involved — as I discovered pretty soon after I got married. One simple glance speaks volumes.

The dividing line between animal and human persuasion gets even more blurred when we consider just how much of the human variety is, like its animalistic counterpart, instinctive. The secret of good advertising often lies not in its appeal to our rational, cognitive faculties, but in its ability to get straight through to the emotion centers of our brains: primal, ancient structures that we not only share with but actually inherit from animals.

I remember when I was a child local town planners, reporters, and crash scene investigators being completely bamboozled by a sudden spate of accidents that had, seemingly overnight, begun to occur at a busy, though previously unremarkable, road junction. A week or so later, the local paper ran a story on its front cover. It featured a bunch of blokes from the council removing a twenty-foot billboard of a curvy, scantily clad blonde from a prominent position nearby.

Sex sells, always has. Even the *word* "sex" sells. In fact, research conducted in 2001 revealed that "sex" appeared on 45 percent of all *Cosmopolitan* and *Glamour* front covers. That simple combination of letters — SEX — acts as a powerful, eye-catching, interest-grabbing, money-spinning key stimulus.

Take, for instance, this clever little flier for an estate agent that came through my front door not so long ago, shown in Figure 1.2.

Figure 1.2 — Into cross-dressing? Then buy a house from us.

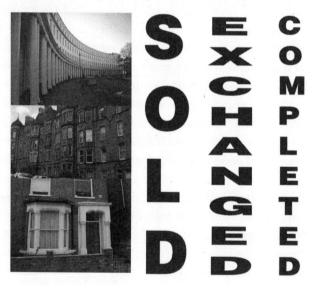

Cheeky, huh?

Of course, marketing supremos and other captains of industry are constantly bombarding us with sneaky, subradar key stimuli. In the relentless campaign for that most lucrative copy space of all — the one between our ears — the deployment of the key stimulus is the psychological equivalent of using a nerve agent. Take the picture of Marilyn Monroe in Figure 1.3.

Figure 1.3 — Nice guitar.

Notice anything strange about it? What about the waist? Does it appear, perhaps, a bit too "hour-glassy"? Images like this, in which the model — either through sheer biological good fortune, überzealous corsetry, or the odd dab of airbrush here and there — exhibits inordinately evocative features, are found all over the place in society (at which point I should explain that this diabolical state of affairs is as distressing to us guys as it is to you girls). And why? Because they sell. But a more pertinent question than "why" is "how?" How do they sell? What is it, exactly, about Marilyn Monroe's midriff in this photograph that gets us so excited? Actually, the answer to this question is simple. What we have here is a biological caricature — a Bell Frog with a megaphone. Or, to put it another way, a "synthetic" key stimulus. Let me explain.

Let us, for a moment, consider Herring Gulls. Herring Gull chicks instinctively respond to a small red spot located on the lower bill of the adult female. Pecking at this spot will result in the adult regurgitating food — the red spot, in other words, constituting a key stimulus. But what exactly is it about this stimulus that makes it "key"? Research has indicated five major factors. By presenting the chick with different models of beak, it's been shown, for instance, that variations in the color of both head and bill are actually of little significance. On the other hand, the red spot itself, narrowness of the bill, movement, low positioning of the head, and a downward pointing of the bill are all essential in generating a response. In fact, so integral to the response are these five core components that a refined, synthetic representation — what is known as a *supernormal* set of stimuli — does the job even better. A thin brown stick with three red stripes near the tip, when moved in a low position, elicits, over and above its original Darwinian prototype, not just a positive response but an *enhanced* positive response. In other words, it pushes the Herring Gull's pecking buttons even harder.

Well, here's the deal.

Precisely these same processes of persuasion at work on Herring Gulls also work on humans — for exactly the same reasons, and by exactly the same mechanism. Supertoned tits and bums, genetically modified lips, six-packs chiseled out of granite, and legs that go on to infinity . . . all of these artifacts are the human sexual equivalents to those three red stripes and that thin brown stick. They are caricatures — quite literally — of the "red-spotted" sexual stimuli that might, at one time or another, have first "caught our eye." And so our responses to them are enhanced.

Figure 1.4 — I'm not dumb, I'm just drawn that way.

WINNING HANDS DOWN

Fortunately for Herring Gulls, the commercial deployment of the key stimulus remains exclusive to humans. Yet it's not just on a corporate level that we're susceptible to this kind of influence. Flashes of the ancient — when persuasion was made of biology rather than psychology — may also be glimpsed in simple, everyday behavior. And when they occur they are dazzling.

I'd been told about Marco Mancini by a friend of a friend at a party. She had worked with him, once, at the Job Centre before handing in her notice and going to live by the sea. She had left, in fact, after only a couple of months — struggling, as many had before her, to keep up the payments on her sanity. Four times, one week, the fire extinguisher bounced off the wall. Not to extinguish fires, but rather to stoke them up, catapulting against the cast-iron security grille that had separated her workstation from the waiting area. Then someone pulled out a gun.

Marco, she said, was different. And a lot of it was in the way that he spoke to people. While everyone else cowered behind plate glass, Marco worked face to face — doing everything out in the open. He always had some coffee on the go. And his desk was right in the middle, where anyone and everyone could see him. That, to her, seemed reckless in the extreme. Insane, even. And, I had to admit, I agreed. But that was the funny thing. Despite all the trouble — and there was, I was told, a lot of it — in the two and a half years that Marco had been at the Job Centre, there wasn't a single recorded instance of him ever having been attacked. Not one.

But there was something else about him, too. It wasn't so much the *way* he talked to people, it was also . . . no, she shook her head. But once people came in to contact with him they seemed to just . . . chill out. As if a switch had flicked or something. Nobody knew why, but everyone had noticed it. Maybe he was crazy, they said. And other crazy people picked up on it.

I was surprised by Marco when I met him. I had expected . . . not sure, really. De Niro in *Heat*? Pacino in *Scent of a Woman*? But I was confronted instead by a trendy, urban Jesus who looked like he worked in a juice bar.

"So, Marco," I said. "In the two and a half years that you've worked at the Job Centre you've been trouble free. What's the secret?"

The secret, it turned out, was surprisingly simple. He sat on his hands. That, plus there was something going on with the chairs. The one for the client facing his desk was adjusted just that little bit higher than his own, so people could literally talk down to him while he listened. Oh, and one other thing. Once things had calmed down a little and the worst of it was over, he would look them in the eye, these angry, crazy people, and smile. And he would touch them, once, on the arm.

"I never forgot something that happened to me when I was ten," Marco told me. "There was this kid at school and he had said something to the teacher about me and I was angry. *Really* angry. I went out looking for him in the playground, and when I found him I was going to beat the shit out of him. And then, when I *did* find him, all I did was shout. And then I shut up.

"It was something about the way he was sitting. He was sitting low down, on a wall, on his hands. I mean, how can you hit someone who's sitting on their hands? It's like shooting someone in cold blood. How can they defend themselves? Plus he had his head down all the time I was shouting, and then he sort of looked straight up at me, still sitting on his hands. It was like he was saying: OK, well, here I am. Hit me if you want. And I couldn't. Somehow I just couldn't. So I left. I walked away."

Such an extraordinary feat of knife-edge persuasive genius should not be undertaken lightly. As well as making the right kind of moves, you need also — if you aspire to be the kind of split-second persuader that Marco Mancini clearly is — to display the right kind of qualities: first and foremost, the confidence and empathy we touched upon briefly in the Introduction (and which, in animal form, we revisited earlier in this chapter). But the moves, nonetheless, are still important — and here's where it gets interesting. On closer inspection, the anatomy of Marco's approach bears a striking similarity to the principles of animal appeasement: symbolic, ritualistic gestures aimed at defusing conflict and "talking your way out of trouble." When escape is not on the menu, and you are.

Take, for example, the thing with the chairs: one being higher than the other. If mimicry is empathy taken straight, then the primordial power of an appeasement key stimulus lies wholly in the art of surprise. Incongruity. Or, as Darwin puts it in *The Expression of the Emotions in Man and Animals*, the "principle of antithesis." A subordinate baboon — regardless of gender — will turn its back on an aggressor and present itself in the mating posture (pseudocopulation). OK, the unfortunate subordinate might, on occasion, actually find itself mounted by the dominant party — mercifully, only briefly—but more often than not the gesture, the *antithesis* of aggression, is accepted as submissive and the subordinate is granted clemency.

Then there's the sitting on the hands. Recent work on crayfish goes one step further than that on baboons—and suggests that appeasement might even be a superior strategy to dominance. When male crayfish compete for female mates, they show each other who's boss by flipping their rivals over onto their backs and then assuming a mating posture. The subordinate animal has one of two choices. It may, on the one hand, offer resistance. Or, on the other, antithetically, take up the receptive female position in submission. Fadi Issa and Donald Edwards of Georgia State University have discovered, much to the delight of the more metrosexual members of the crayfish population, that kicking back and letting the macho types get on with it actually pays dividends. After twenty-four hours of pairing off, half of the resisters were killed while all of the submitters survived.

Taking it lying down, or in Marco's case sitting down, clearly has its advantages.

STOOPING TO CONQUER

The knowledge that we have of key stimuli, of how they work, and of the powerful influence they exert in the animal kingdom, allows us, as we just saw with Marco, to turn them to our own advantage. Just as the largest, sturdiest buildings can be made to collapse under their own weight by the careful placing of explosives, so even the most intractable of problems can be dismantled by a few carefully bestowed words and gestures. Throughout history the great persuaders have known this.

In the Gospel of St. John, for example, Jesus finds himself cornered. The Pharisees present him with a woman accused of adultery, and petition him for advice.

"Master," they say. "This woman was taken in adultery, in the very act. Now Moses in the law commanded us, that such should be stoned. But what sayest thou?"

The Pharisees, of course, aren't really interested in Jesus's moral take on the matter. And Jesus knows it. Instead, their motives are altogether less salubrious. What they're actually trying to do is get him embroiled in a legal wrangle. According to Mosaic law, the woman, as the scribes correctly point out, should be stoned. No problem there—under normal circumstances. But with Palestine now under Roman occupation, things have changed. Mosaic Law has ceded to Roman law — and if Jesus upholds the former over the latter, he leaves himself open to the inevitable charge of incitement. But that's the least of his worries. Conversely, if he decrees that the woman should not be stoned, he stands accused of precisely the opposite charge — turning his back on the ancient traditions of his forefathers. And that's no picnic either.

A crowd has gathered, and tensions are running high. Getting out of this, it would seem, is a pretty hard task even for the smoothest of smooth talkers — let alone an itinerant carpenter with no rhetorical training whatsoever. What happened next is described thus:

> This they said, tempting him, that they might have to accuse him. But Jesus *stooped down,* and with his finger wrote on the ground, as though he heard them not. So when they continued asking him, he *lifted up himself,* and said unto them, He that is without sin among you, let him first cast a stone at her. And again he *stooped down,* and wrote on the ground. And they which heard it, being convicted by their own conscience, went out one by one, beginning at the eldest, even unto the last: and Jesus was left alone, and the woman standing in the midst. (John 8:6–9; author's emphases)

This passage from the Gospel of St. John is unique. It's the only recorded occasion in the entire New Testament during which Jesus writes anything. Speculation is rife among Biblical scholars as to what those words might have been. The sins of the woman's accusers? Their names, perhaps? They will, of course, forever remain a mystery. But from a psychological perspective, precisely why Jesus should feel the need to write anything at such a moment constitutes an even greater conundrum.

It doesn't make sense.

Unless, that is, he had something up his sleeve. Might the words themselves have been a smokescreen? The significance of his actions lie less in the writing itself — and more in the act of producing it?

Let's take another look at Jesus's body language during his encounter with the Pharisees. The exchange, in fact, comprises three distinct phases. On first being challenged, what is his initial reaction? Well, we note from the text that he immediately "stoops down" (antithesis: incongruity: appeasement). Then, when the elders persist in their sophistry, he "lifts himself" back up again to deliver his famous riposte (confidence: assertiveness) — before reverting to a stooping posture and resuming a pose of appeasement.

It's a well-crafted move aimed at shifting and stealing momentum.

Sure, Jesus certainly has a great line in "casting the first stone." And, what's more, he almost certainly knew it: it's one of the finest examples of split-second persuasion I've ever come across. But he does, however, still have a problem. At the end of the day, no matter how great a line it may or may not be, no matter how insightful the argument, it still delivers a challenge to the Pharisees. And could, despite its genius, have seriously pissed them off.

An eventuality, no doubt, of which Jesus was well aware.

And which explains, contrary to theological conjecture, why he didn't just speak in the one language, he spoke in *two*. One modern, phonemic, opaque. One ancient, silent, profound.

FIRE AND RESCUE

Marco Mancini and Jesus have little in common. True, Marco did look a bit like Jesus when I met him. But that, I would guess, is where the similarity ends. Marco first learned the secrets of split-second persuasion in the school playground. Jesus . . . who knows? The point is that one doesn't need supernatural powers to excel at persuasion like this. The ability lies within all of us. But unlike our animal brethren, *we* have to work at it.

Neither, of course, is such influence restricted to flashpoints. OK, it may, from time to time, help get us off a ticket. Or the end of somebody's fists. But it can also help us in other ways as well. Think about it. The more you can say without actually having to say it gives you one hell of an advantage no matter what situation you're in.

Take business. Research has shown that top salespeople often lean slightly forward toward their clients when doing deals — a double whammy signifying not only empathy (through increased proximity) but also a sneaky subservience.

Or parenting. Next time you find yourself having to lay down the law to a wayward six-year-old, try laying it *up* instead. Rather than towering over them, draw them up close, crouch down next to them, and then — in as calm a tone as possible (I know, easier said than done) — say what you have to say.

Bringing yourself down to someone's level like this often speaks volumes. Remember Churchill and the dinner party thief from the Introduction? What you are saying (without actually having to say it) is this: "Look — it's not just *you* that's in the shit here. It's *both of us.* So why don't we see if we can't work as a team from now on. Deal?"

Here's Winston again — up to his old tricks.

In the summer of 1941, Flight Sergeant James Allen Ward was awarded a Victoria Cross for clambering onto the wing of his Wellington bomber and — while flying 13,000 feet above the Zuider Zee — extinguishing a fire in the starboard engine. He was secured at the time by just a single rope tied around his waist.

Some time later, Churchill summoned the shy, swashbuckling New Zealander to Number 10 Downing Street to congratulate him on his exploits.

They got off to a shaky start.

When the fearless, daredevil airman — tongue-tied in the presence of the great man — found himself completely unable to field even the simplest of questions put to him, Churchill tried something different.

"You must feel very humble and awkward in my presence," he began.

"Yes, sir," stammered Ward. "I do."

"Then you can imagine," said Churchill, "how humble and awkward I feel in yours."

SUMMARY

In this chapter we've looked at the ancestry of influence. How persuasion was done before the advent of language, and how it's still being done in the animal kingdom today. The conclusions we've come to are stark. With the arrival of language, and the rise of the neocortex, persuasion, rather than becoming more effective, has actually become less so. When it comes to persuasion, animals do better than we do.

The secret of persuasion in the animal world is thrift. In animals, the basic units of influence are what ethologists call key stimuli — persuasion silver bullets which, when fired by one member of a species at another, generate instinctive, preprogrammed response sets. These silver bullets — innate, immediate, and incisive — resolve situations quickly, and with a minimum of cognitive fallout. With humans, however, it's different. Wedged between us and the expediency of instinct is an ozone layer of consciousness — which language, our influence tool of choice, often finds hard to penetrate. Only the really special make it through.

The question, of course, is how to fashion such influence. Are all of us capable of hitting these persuasion sweet spots? Or is it just the preserve of a handful of influence elite?

You may find the answer surprising. Each of us is born under the star of persuasion genius. But as we get older its luster slowly wanes.

2

Fetal Attraction

A Houston lady just told me that her friend heard a crying baby on the porch the night before last so she called the police because it was late and she thought it was weird. The police told her: "Whatever you do, do not open the door." The lady then said she thought that the baby had crawled near a window and she was worried that it would crawl into the street and get run over. The policeman said, "We already have a unit on the way. Whatever you do, do not open the door." He told her that they thought a serial killer had a baby's cry recorded and was using it to coax women out of their homes, thinking that someone had dropped off a baby. He said they had not verified it but had had several calls from women saying that they heard babies' cries outside their doors when they were home alone at night.

The helpless cry of a human baby is not weak and ineffective and archaic. It is the most profound and powerful force in nature. Until a father and mother first hear it, parenthood lies dormant in them. . . . The [infant's] cry is not into space but down into the profound of human love and pity.

— JONATHAN HANAGHAN, *Society, Evolution, and Revelation*

PERSUASION PRODIGY

I am sitting in a café in South London about to meet a man who, a couple of years ago, wouldn't have been seen dead in a place like this. There are plenty of other joints he might have been seen dead in around here. But this place? With its Fair Trade mochas and Ugg boots? No. Here he comes now. Although I've never seen him before I'm pretty certain it's him. He's tall — around six feet three inches. Late twenties. And has a tan. Not the kind of tan you get in Greece but the kind of park tan that winos get from hanging around on benches for long periods. His name is Daryl.

Daryl has spotted me — I'm right, it *is* him — and he's coming over. The first thing I notice about him is the shaking. Somewhere in his brain the weather's turned bad and some power lines are down. Then there's the scar. And the amateur tattoos. And the sports bag, which he dumps against my leg. What's in there? I think.

A couple of years ago Daryl was part of a low-level crime syndicate that operated around these parts. He was high on crack for most of that time. And you name it, he did it. Everything from housebreaking to robbery to obtaining false passports.

Then one day it all went right.

Walking through a car park one Saturday afternoon, he saw a woman loading some shopping into a car. He pulled out a knife and approached her from behind. But when she turned round, Daryl got a surprise. Clasped between her arm and her body was a newborn baby. He froze. She froze. They *all* froze. The baby just stared at him and Daryl stared back. Then the woman screamed, and he dropped the knife and ran. Then he went into rehab.

"I just don't know what happened," he tells me as I sip my triple, no foam soya latte, with one pump of sugar-free vanilla, swirled. "But something fucking did. The kid was a real shock. The way it looked at me . . . it was, like, whoah! I can't be dealing with this shit. I never wanted to hurt nobody. I just wanted money. You know, for smack and that. Once, like, that kid would have been me. How did it come to this? What happened to the kid *I* once was? I thought."

MISSION IMPOSSIBLE

The newborn baby is a persuasion machine. There's no other way of putting it. The ability of the neonate to impose its will on others, to get its own way, to twist us round its tiny little fingers, is second to none. It has the techniques of social influence down to a fine art.

The human response to infants is pretty much universal across any demographic you may care to mention. Culture, age, sex — you name it, the reaction is almost identical. Take age. Studies have shown that infants as young as four months look longer at pictures of infant faces than they do at those of older children or adults. And that, by eighteen months, this preference for infant faces is accompanied not only by increased smiling, but by gesturing and vocalizing, too. More surprising still, perhaps, is that this doesn't just happen in humans. From the age of about two months on, rhesus monkeys reared in isolation show a similar preference for pictures depicting infant monkeys over those depicting adult monkeys. The appeal of the neonate is set in neural stone.

These perceptual biases lurk deep within the brain. Research conducted by neuroscientist Morten Kringelbach at the University of Oxford reveals that when pictures of newborns are viewed under MEG (magnetoencephalography — an imaging technique that monitors brain activation in milliseconds), the area of the brain that encodes for rewarding stimuli — the medial orbitofrontal cortex — responds almost immediately: within one-seventh of a second of the images first appearing.

Our brains, says Kringelbach, have a built-in propensity to "tag" infant faces as special.

Precisely why the newborn should be so persuasive is not exactly rocket science. Like many things in life it essentially boils down to marketing. Chimpanzee babies fuss. Seagull chicks squawk. Burying beetle larvae tap their parents' legs. Throughout the animal kingdom, newborns display a consummate knack for engaging their parents' attention — exhibiting an eclectic array of ingenious key stimuli to elicit nurturance and inhibit adult aggression.

Such advertising is of the utmost importance. There was a time (not that we remember it too well) when each of us — alone and without backup — first showed up for life. It was a pretty risky maneuver. Consider the enormity of the initial challenge that faced us. Somehow, from the very first moment we came into this world, we had to influence those around us — without thought, without language, without control of even the most basic of bodily functions — to take care of us. Somehow we had to persuade them it was worth it.

Nowadays, of course, we take all this for granted. Because we made it. Not, I should add, through any recourse of our own (if we'd had anything to do with it, who knows what might have happened?) but through the genius of natural selection. Through power of biological attorney, natural selection saw us on our way: equipping us not just with one key attribute — the ability to raise hell — but with three. Three key stimuli of social influence, fitted as standard, we find tantalizingly irresistible:

- The ability to cry with brilliant acoustical efficiency
- A devilish cuteness (which, for those fortunate enough to hang on to it, also works in later life as well)
- A hypnotic capacity to make eye contact

Persuasion, no matter what its shape or form, comes no more incisive than that.

In this chapter we continue our quest for the origins of social influence, the primeval lineage of changing the minds of others, by looking a little more closely at these three key stimuli of neonatal influence. What precisely is it about crying that makes it so special? And what are the features of newborn infant faces that cast such a spell on us?

WIRED FOR SOUND

Instant influence isn't something we come across all that often in the sphere of human interaction. We saw as much in the previous chapter. With us, unlike animals, things usually take time. This is largely because we have brains big enough to drive a bus through. We have the ability to learn. To reflect. To decide. And then to talk about it afterward. But dotted along the bus route lurk vestiges of the past: ancient, disused stations that can sometimes come back to life. Certain modes of communication, certain means of interaction, can, to this day, make us do stuff without our even thinking, thanks to their overarching importance in our evolutionary history. There are times in our lives when our brains give reason the elbow.

In 1998, the Pentagon commissioned Pam Dalton of the Monell Chemical Senses Center in Philadelphia to come up with something unusual. Intrigued by the somewhat Python-esque notion of controlling public order by smell, the U.S. government put Dalton in charge of the world's most dangerous chemistry experiment. She was to create, for the first time in history, a universally repellent bouquet. Could there, wondered senior officials in the U.S. Defense Department, exist something so malodorous as to disperse a rampaging mob on a moment's exposure?

Dalton found out that there could.

In fact, she discovered there wasn't just one such infusion but two. A pair of equally odiferous concoctions that transcended not only all known individual differences, but all known cultural divides as well: the aptly named U.S. Government Standard Bathroom Malodor (which comprises — surprise surprise — the concentrated stench of human feces), and the equally apt, though somewhat elliptically named, Who Me? (a hideous collection of sulfur molecules which simulates the mephitic aroma of rotting carcasses and spoilt food). Results, one might say, not to be sniffed at.

Technological advancement has often proved the wellspring of inspiration — and anyone who has ever had to toss and turn on a windy night while their neighbor's car alarm goes off in the driveway next door, or who has been driven to distraction by the idiot opposite's customized ringtone, will no doubt be wondering if similar research has not been conducted on sound. It has. Snoring, squabbling, coughing, and farting are among the primary contenders here.

To many, this may come as something of a surprise. Compared to the racket of a three-pronged garden pitchfork being scraped across slate — revealed by a study in the 1980s as being the most likely candidate to elicit disgruntlement — these more "organically" flavored sounds appear fairly innocuous. But as Trevor Cox of the University of Salford's Acoustics Research Centre points out, the inference of irritation from the physics of a sound wave alone is not as straightforward as first it might appear. There's psychology in the mix as well.

"The boom-boom of your neighbor's hi-fi isn't so annoying if you are going to join the party later," Cox says. And he's right. In short, as is the case with many potential stressors, the amount of stress that actually occurs depends on how much control the recipients have, or perceive they have, over their surroundings. "If you have control over the noise, it tends to be less annoying," says Cox. "But if you are fearful of the source, then it usually makes it worse."

British inventor Howard Stapleton has recently put Cox's theory to the test. Quite literally, in the marketplace. His Mosquito device — designed, like Dalton's irresistibly repellent odors, to reduce antisocial behavior by dispersal — emits an irritating, high-pitched warbling sound of such a frequency as to be inaudible to anyone over the age of thirty. The contraption, dubbed an "electromechanical teenager repellent," is currently in use in high streets and shopping malls right across the U.K. and is proving at least as successful as its predecessor in the war on yobbishness: Wagner. Somewhat more muscular than coughing and farting (it weighs in at 85 decibels — about the same as a lawnmower) yet not sufficiently macho as to inflict actual physical harm, Mosquito's main advantage appears to be its nuisance value. Latest reports suggest it's doing well.*

THE CRYING GAME

The normal range of adult human hearing stretches from approximately 40 Hz to 15 kHz. The range covered by the human voice is typically that between 100 Hz and 7 kHz. And then there's the point at which human hearing is at its most acute: around 3.5 kHz. From the standpoint of natural selection, this is interesting. A number of sounds operate at frequencies of around the 3.5 kHz mark (submarine sonar, for example) — sounds specifically developed for scenarios in which a high premium is placed on attentional resources.

Yet there's another sound, pitching in at around 200 Hz to 600 Hz, which has a somewhat longer lineage. And which, of all the acoustical stimuli known to man, places the greatest demands of any on our attention: the crying of a human infant.

An infant's cry is genius made sound — the profoundest thing that can ever happen to an air molecule. It operates at two fundamental, though not unrelated, levels of influence: physiological and psychological. In common with other alarm and emergency signals, its acoustical properties evolved, quite literally, under cover of darkness — eliciting attention and conveying position to caregivers while at the same time minimizing location cues to predators. (The high-pitched frequencies of infant vocalizations aren't as aerodynamic as low-pitched frequencies — favoring nearby members of one's own species over distant roaming killers.)

But there's more to the cry of the neonate than just the provision of localization cues. In addition to its base-station benefits, its tonal, graded signal elicits an instinctive physiological response set in caregivers: cardiac deceleration followed by rapid acceleration (associated with imminent action or intervention), together with elevated breast temperature and a milk-letdown reflex — which makes the breasts feel heavy and stimulates the mother to feed.

---

*Using the same technology, Stapleton has also come up with the "silent ringtone" — a phone audible to teens but not to their teachers. That should make lessons more interesting.

Back in the days of our ancestors, an infant's cry was the ultimate 911 call. Pizza delivery, too.

Paradoxically, however, infant crying "grates." Though its auditory representation falls short of the range of "maximum acoustic unpleasantness"* (its high-pitched tone is high enough not to carry, but sufficiently low so as not to engender aggression), the cry of the neonate is up there on virtually everyone's list of aversive acoustic stimuli — men and women, parents and children alike — evoking anxiety, distress, and overwhelming urges to "help."

The equivalent, in sound and empathy, of U.S. Government Standard Bathroom Malodor.

In 2007, Kerstin Sander of the Leibniz Institute for Neurobiology in Germany demonstrated precisely how deep an infant's cry goes. Sander played recordings of four different cries to a group of eighteen adults (nine men, nine women) while they underwent fMRI.[†] She then scrambled the cries (splitting each recording into 150-millisecond segments), recombined the fragments, and compared what happened next. Would patterns of brain activity remain constant across both crying conditions? Sander wanted to know. Or would the scrambling make a difference?

Her findings bore testament to the choreographic genius of natural selection. Results revealed a dramatic increase in both amygdala activity (the part of the brain which processes emotion) and anterior cingulate cortex activity (the part of the brain sensitive to anomaly) when the *real* cries were played as opposed to the scrambled segments. But more so for women than for men — a pattern, Sander suggests, which may reflect a specific neural predisposition in women to respond to preverbal infant vocalization (see Figure 2.1).

And when Sander took a subset of her participants and compared amygdala activity for natural infant crying against natural adult crying, she *really* got a shock. Here, the increase for infants was even more pronounced: 900 percent. An infant's cry, contrary to outward appearance, isn't quite as simple as it seems.

---

*Around 2,500–5,500 Hz, with temporal modulations in the region of 1–16 Hz.
[†]fMRI, or functional magnetic resonance imaging, is a technique measuring blood-oxygen levels in the brain, thereby allowing researchers to determine which areas are most active at any given time.

Figure 2.1 — (Left) Approximate areas of female brain activity on hearing an adult crying. Shaded sections indicate areas of increased brain activity. (Middle) Approximate areas of female brain activity on hearing an infant crying. (Right) Approximate areas of male brain activity on hearing an infant crying.

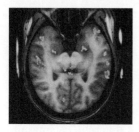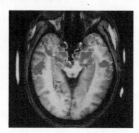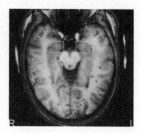

WHEN THE WRONG NOTE HITS THE RIGHT NOTE

And nor, it would appear, as uniform, either. Additional research has probed a little deeper — and revealed that although preverbal infant vocalizations do indeed increase amygdala activation, it is sudden, dramatic, and unexpected changes in crying pitch — known as "gliding" or "vibrato" — that convey the most emotion, and elicit in caregivers the most powerful affective responses. Moreover, it's precisely these same unexpected shifts that move us so intensely when it comes to music: that send a shiver up our spines, that create that inimitable "tingle factor." It's not the predictable resolution of chords that loosens the screw top on our emotions. Nor, for that matter, in comedy, is it the expected that makes us laugh. It's the dotty, ecstatic ignominy of being mistaken.

As an example, take the following. Paul Rozin and his colleagues at the University of Pennsylvania have drawn attention to a common pattern that exists in humor — what they call the AAB pattern. We all know what it is:

> (A1) Some men are about to be executed. The guard brings the first man forward, and the executioner asks if he has any last requests. He says no, and the executioner shouts, "Ready! Aim!"
>
> Suddenly the man yells, "Earthquake!"
>
> Everyone is startled and looks around. In all the confusion, the first man escapes.
>
> (A2) The guard brings the second man forward, and the executioner asks if he has any last requests. He says no, and the executioner shouts, "Ready! Aim!"

Suddenly the man yells, "Tornado!"

Everyone is startled and looks around. In all the confusion, the second man escapes.

(B) The last man has it all figured out. The guard brings him forward, and the executioner asks if he has any last requests. He says no, and the executioner shouts, "Ready! Aim!"

And the last man yells, "Fire!"

In this particular joke (I've got others), the violation — B — involves a rival interpretation of the final word. The expectation is one of another disaster word. But in reality, when it comes, it has a different, calamitous, and radically pertinent undertone. Perhaps less well known, however, is the AAB pattern in music.

Figure 2.2 — The AAB structure in the initial theme from Mozart, Piano Sonata in A Major, K. 331, I (Andante grazioso), mm. 1–4.

Here, we see the original five-note motif (A1) repeated a tone lower (A2), then begin again another step lower still (B). Only in this third instance, it changes form to a different note sequence entirely. Such "script reversals" are common within a wide variety of musical genres from classical and contemporary through to Broadway and jazz. As well as, needless to say, forming the basis of countless "Englishman, Irishman, Scotsman" and "Priest, Minister, Rabbi" jokes.

Does such manifest incongruity, such conspicuous violation of expectation, constitute a universal law of persuasion? Quite possibly. It's certainly a component of what Darwin had in mind with his principle of antithesis — the script reversal so integral to appeasement displays in animals. And, as we saw in the previous chapter, humans, too.

"Music," writes V. S. Ramachandran of the University of California's Center for Brain and Cognition, "may involve generating peak shifts in certain primitive, passionate primate vocalizations such as a separation cry; the emotional response to such sounds may be partially hardwired in our brains."

David Huron, in his book *Sweet Anticipation: Music and the Psychology of Expectation,* goes one step further.

"Forming expectations," argues Huron, "is what humans and other animals do to survive; only by predicting the future can we be ready for it. And because the brain ensures that accurate prediction is rewarded, we feel good when we are proved right. The link between prediction and reward causes us to constantly seek out structure and predict how events will unfold. As a temporally evolving texture, music is a superstimulus for such predictions."

In other words when expectations are violated, our brains (more specifically, areas such as the anterior cingulate cortex and parts of the temporoparietal junction) are moved to restore homeostasis. To counteract the aversive emotion that accompanies such violation. In the arts — music and comedy, for instance — such aversive emotion is all part of the deal. From the comfort of our armchair, or the safety of the dress circle, we place ourselves quite willingly in the hands of the performer.

But in other areas of life we're not so charitably disposed.

When an event or stimulus defies prediction we're forced to do something about it: either to discredit or to eradicate it. Or else to rethink our position. Which is why it's virtually impossible — especially for caregivers — to ignore a baby's cry. Not only does the sound itself generate aversive feelings, so too do deep, essential elements of its structure.

BEAUTY AND THE BEST

You can spot them a mile off. In fact they're probably visible from outer space — those guys in the shopping mall with clipboards. For some reason, it always seems to happen when you're in a hurry. Or have discovered, just moments before, that your house is on fire.

"Can you just spare a couple of minutes to answer a few questions . . ."

Many of us have evolved elaborate strategies in the face of such a predicament. Persistent coughing. Mobile phones springing suddenly into life.

And the spontaneous appearance of imaginary acquaintances on the other side of the road. All of which count for nothing, of course, should the person holding the clipboard be a fit, attractive blonde. In which case, far from plucking out our own eyeballs in the desire to avoid visual contact, most of us form a line.

To a social psychologist, one well versed in the vicissitudes of interpersonal attraction, such an occurrence will come as no great surprise. It's a well-known fact that good-looking people accrue more signatures on petitions than their face-like-a-slapped-arse counterparts, and that charity stands manned by good-looking volunteers generate increased takings. In the law courts, too, attractiveness plays its part. Good-looking defendants are less likely to be found guilty than those of average appearance. And, in the eventuality that they are, receive lighter sentences. Good-looking people are good.*

Flick through the pages of a dozen pop psychology mags and you'll be hard pressed not to encounter a hundred such claims as these. Good-looking people are better at such-and-such. Average-looking people do worse. Yeah, yeah. But where's the evidence? Mark Snyder at the University of Minnesota conducted a study in which male students were presented with an information pack containing details relating to a fellow female student (in reality, an associate of the researchers). These details included a mocked-up photograph of the student which had been designated by the researchers as either attractive or unattractive. Under the pretence of discussing coursework requirements, the researchers then contrived a ten-minute telephone conversation between the participants and their fellow "students" (in all cases, the same person) and observed the way in which the participants interacted. Would attractiveness impact on telephone manner? they wondered.

The answer was yes. Big time. Participants who believed that the person they were talking to was attractive responded to her in a warmer, more positive fashion than those who believed she was unattractive. Furthermore,

---

*Such findings may be explained by something known as the *halo effect*: the presence of one or two discrete positive traits — including physical attractiveness — which invoke, in turn, a generic aura of goodness, expertise, honesty, or some other inferred superlative. Interestingly, it's not the case for all crimes that good-looking defendants are less likely to be found guilty. There is, in fact, one type of crime where the likelihood is actually *greater*. Can you think what that kind of crime might be? The answer is given at the end of the chapter.

when asked, prior to the conversation, to record their initial impressions of the student, expectations clearly differed on the basis of attractiveness. The participants who were presented with the attractive photograph anticipated interacting with an outgoing, humorous, and socially skilled individual. The participants who got the ugly photo, didn't.*

BRAIN TEASER

In 2007, in a study involving lap dancers, strip clubs, and sex pheromones, the evolutionary psychologist Geoffrey Miller uncovered another predictor of attraction — this time in the adult entertainment industry. Over a period of a couple of months, Miller and his coinvestigators Joshua Tybur and Brent Jordan took 5,300 erotic performers (yes, that *is* the correct number) and divided them into three groups: those who were ovulating, those who were menstruating, and those who were in between. The question was simple. Which of the three groups would make the most money by the end of their five-hour shift?

According to the precepts of evolutionary psychology, it should have been the girls who were ovulating. In the event of anyone getting lucky, this was the group most likely to conceive. And they did. Make more money, that is. True to form, punters found the ovulating dancers more appealing. And, in a show of appreciation, coughed up. In fact, the results of the experiment couldn't have turned out better. On average, the ovulators made $325 in tips. The menstruators $185. And those in between, $260.

Miller's study is intriguing for a number of reasons. But the main one is this. Much of the time we are as able to put into words why we find someone attractive as we are able to put into words why we like the particular type of music that we do. Sure, we might invoke specific aspects of that music such as the rhythm or the harmony, but the questions still remain. Why *that* particular rhythm? Why *that* harmony?

Let's, in a manner of speaking, turn the tables on Miller for a moment and look at a predictor of male attractiveness. In Figure 2.3 (overleaf), which face do you think is the better-looking: The one on the left? Or the one on the right?

---

*Just in case you were wondering, it works the other way round as well. Susan Andersen at New York University and Sandra Bem at Cornell University manipulated the role-playing scenario so that the participants were female and the bogus fellow student was male. It made no difference.

Figure 2.3—Spot the difference.

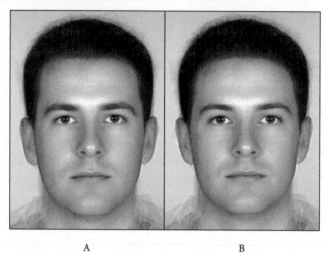

A   B

Generally speaking, most women go for the face on the right. Unless they are ovulating, in which case they go for the other one. But ovulating or not, they just can't put their finger on why. David Perrett, on the other hand, of St. Andrews University in Scotland, knows exactly where to put his finger: on imperceptible cues of gender enhancement. In a jaw-mashing blow to Arnold Schwarzenegger and his pals, Perrett has found that women, on average, actually prefer men's faces when they are made just that little bit more like their own. When they are feminized, in other words. Here, the face on the right has been feminized by 30 percent — the optimal amount to maximize attractiveness. The jawbone, you will notice, has been rounded and smoothed out. And the forehead and eye regions softened. But during ovulation the trend mysteriously reverses. For women who are ovulating, it's actually masculine facial features that prove the bigger turn-on. Stronger, robuster, they ruggedly allude to greater immunological competence — a heritable resistance to disease — and faces that embody a more macho physiognomy assume a subtly heightened significance (see Figure 2.4a).

On the other hand, however, there's one marker of attractiveness that pretty much suckers all of us; that cuts through the crap of conscious-unconscious processing and, just like the good-looking student in the telephone experiment, plays upon our better natures. This is the mark of the baby face.

Figure 2.4a—During ovulation, women generally prefer the characteristics of more masculine faces such as that of Bruce Willis (left) over the more feminine features of, say, Leonardo DiCaprio (right). The enhanced appeal of faces such as that of British actor Robert Pattinson (Figure 2.4b) lies in their combination of both masculine *and* feminine attributes—note the refined jawline, full lips, and low, prominent eyebrows.

Figure 2.4b

YOU'VE GOT THE CUTEST LITTLE BABY FACE

In 1943, in his classic paper "The Innate Forms of Potential Experience," the Austrian ethologist Konrad Lorenz put forward a radical notion. Human beings, he proposed, are equipped with a built-in preference for infant facial features over the facial features of adults. The primary reason for this preference, he contended, centers on caregiving. An innate perceptual bias toward the faces of neonates increases the incentive to protect and look after vulnerable members of the species. To illustrate his point, Lorenz produced a series of silhouettes of both human and animal infant faces which delineated a distinct subset of features — kindchenschema (or baby schema) as he called them — that were common to both: a soft, round cranial shape; a wide curved forehead; large round eyes; and round protruding cheeks (see Figure 2.5). These, he argued, instinctively hot-wired attraction and opened the door to compassion. They were the human key stimuli of nurturance.

Figure 2.5—Cross-species similarity in babyish versus mature facial characteristics.

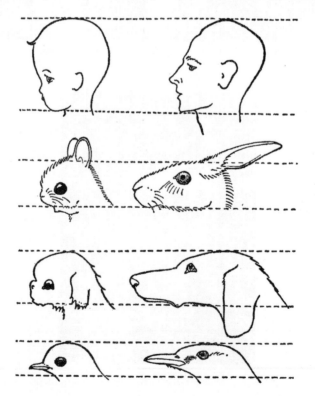

Subsequent research into the science of facial attractiveness has revealed several more of these *kindchenschema*: a small chin; a small, short nose; and the relatively "low" positioning of the eyes, nose, and mouth. All of which, it would seem, are the epitome of "cute." In fact, so powerful are such stimuli in signifying immaturity that they even transfer to random, *inanimate* objects.

Consider, for example, the series of craniofacial profiles illustrated in Figure 2.6. Here, the shape of the head has been sequentially modified using a mathematical transformation that simulates, on cranial geometry, the effects of maturation. Few of us have any difficulty in identifying the direction of increased maturity as proceeding from left to right. Yet here's the deal. Not only do we find it easy to differentiate between a mature and infantile head shape — we also find it easy with mature and infantile *cars*!

Figure 2.6—Changes in craniofacial profile shape with increasing maturity.

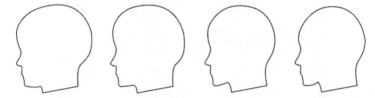

Take a look at Figure 2.7, for instance. The same mathematical function just used to simulate cranial maturation has also been applied to one of the Volkswagen Beetles. Which do you think it is? Which one of the cars do you think features "earlier," and which "later," in the growth transformation stage? Which of the Beetles is *cuter*?

Figure 2.7—Babyish versus mature craniofacial characteristics in motor vehicles.

FACING UP TO RESPONSIBILITY

In 2009, Melanie Glocker of the Institute of Neural and Behavioral Biology at the University of Munster conducted an experiment to test Lorenz's theory. Do we really find *kindchenschema* more appealing? And, if so, how are such preferences reflected in the brain? Using a similar technique to that of Morten Kringelbach, Glocker presented participants with pictures of newborn infants, while they underwent fMRI. Only this time she went a step further. Whereas in Kringelbach's study the pictures were always true likenesses, Glocker, using a special image-edit program, manipulated the images so that some were more "babyish" than others (see Figure 2.8).

Figure 2.8—Manipulated (low/high) and unmanipulated baby schema.

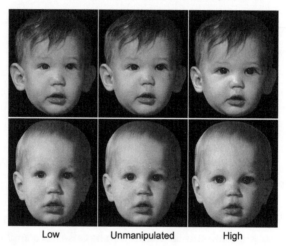

| Low   Unmanipulated   High |

The results turned out exactly as Lorenz would have predicted. Analysis revealed that the higher the index of *kindchenschema* (e.g., the larger the eyes and the rounder and chubbier the face), the greater the activity in participants' nucleus accumbens — the part of the brain, in both humans and animals, that mediates reward. Not only were there *kindchenschema*, Glocker discovered, but super*kindchenschema*, too.

Glocker's adventures deep within the brain have their parallels in everyday life. Imagine you found a wallet on the street. What would you do? Report it to the police? Post it back to the owner? Er . . . hang on to it? Psychologist Richard Wiseman of the University of Hertfordshire posed exactly this question to the people of Edinburgh. Only there was a catch — he did it for real. Wiseman left a bunch of wallets on the streets of the Scottish capital, each containing one of four photographs: a smiling, happy couple; a cute, cuddly puppy; a contented elderly couple; and a beaming, bouncing baby.

Which ones, he wondered, would find their way back to their "owners" most often?

He certainly got his answer. Of the forty wallets of each type that were dropped, 28 percent of the contented elderly couples made it back successfully; 48 percent of the smiling, happy couples; 53 percent of the cute, cuddly puppies; and a whopping 88 percent of the beaming, bouncing babies.

"The baby kicked off a caring feeling in people," says Wiseman — a nurturing instinct toward vulnerable infants that has evolved to safeguard the survival of future generations.

Another study in America evoked a similar kind of protectiveness by pinning a picture of a baby's face to a dartboard. Participants were given six darts and were paid a quarter for every dart that hit the target. Despite the fact that they'd had six "warm up" throws using a face-shaped circle, guess what? Participants were less accurate when aiming at the baby's face than they had been previously.

And it's not just studies with neonates that lend credence to Wiseman's claims. Studies involving adults with "baby faces" also suggest that there's something a little bit special about the features of newborn infants. Sheila Brownlow and Leslie Zebrowitz of Brandeis University conducted a systematic analysis of 150 television commercials. How, they wondered, did content and presenter match up? To find out, they solicited the help of two groups of college students. One group of students read through transcripts of the commercials, rating as they did so the degree of trustworthiness and expertise reflected in each message. The other group viewed recordings — this time rating the faces of the communicators on a scale of facial maturity.* But there was a catch. Crucially, these latter ratings (those for faces) were provided in the absence of sound — thus entailing a "double dissociation" between the information known to each group. The first group got the message without the faces. The second got the faces without the message. How would the two compare?

Results revealed a telltale pattern. On those occasions when the persuasive appeal relied less on expertise (the knowledgeable communication

---

*This scale ranged from baby-faced-ness (associated strongly with trustworthiness and weakly with expertise) on the one hand, to mature-faced-ness (associated strongly with expertise) on the other.

of objective and valid facts) and more on trustworthiness (the sincere and honest endorsement of a product user), it was the actors or actresses with baby faces that tended to front the commercial. In contrast, however, when the flavor of the appeal veered more toward the "factual," so the face of the spokesperson morphed into greater maturity.

And such contours of persuasiveness aren't just found in advertising. In politics, too, they're also well represented. Research has shown that when voters believe a candidate to be acting out of self-interest, it's "honest" baby-faced politicians, as opposed to their more "inscrutable" mature-faced colleagues, who are rated the more persuasive. In contrast, however, when it's expertise that's under the microscope, it's the "shrewd," more mature-faced politicians that are found to be better persuaders.

It's actually rather interesting to compare the faces of different politicians, to see how they fare in the honesty stakes. Back in 2008, a team from the University of Kent got one hundred members of the public to rate a series of faces on a scale of 1 to 5 for how trustworthy they seemed. From the data, they then assimilated a number of features generally associated with honesty: a fuller, more rounded face; a softer, smoother jawline; big, round eyes; and softer eyebrows. Ring any bells? Facial hair was viewed with deep suspicion, but a refined nose and a larger, thinner mouth were also rated positively.

Using a digital image-enhancement program, the team fed in faces of various politicians and examined the difference between how they appeared "normally" — in everyday life — and their trustworthy/untrustworthy avatars. The program wasn't the only thing that raised eyebrows. Take a look, in Figure 2.9a, at the disparity between Gordon Brown's normal facial features and those of his "trusted-up" clone. Then, in Figure 2.9b, do the same for David Cameron.

Figure 2.9a—Digitally manipulated images of Gordon Brown showing (left to right) the original image, trustworthy features, and untrustworthy features.

Figure 2.9b—Digitally manipulated images of David Cameron.

Brown comes off worse because of his "thick eyebrows, wide nose, and the size of his mouth." Cameron, on the other hand, has a "fresh-faced, smooth complexion, wider mouth, and more rounded eye-shape."

Plastic surgery to make you seem more trustworthy? Only a matter of time . . .

Studies like these are just the tip of the iceberg. Researchers have, in fact, uncovered all sorts of differences between baby-faced and mature-faced individuals. Or, more specifically, in the ways we interact with them. In relationships, women are more likely to confide in a baby-faced friend than in one who looks more mature. In the courtroom, baby-faced defendants are more likely to be found guilty of crimes involving negligence than those involving intentional misconduct (the reverse being true for mature-faced individuals). And in the workplace, baby-faced individuals are less likely than their mature-faced colleagues to hold positions of power.

Take the four photographs of military cadets shown in Figure 2.10a. Do you think you can predict, on the basis of appearance alone, how successful each of them will be in their chosen career? Do you think you can tell, from the way they look at the beginning of their military service, the rank they

Figure 2.10a—Portraits taken from *The Howitzer,* 1950.

A   B   C   D

will hold at the end? Have a go. Rank the faces in order putting the one you think will be most successful first, and the one who you think will be least successful last.

How did you get on? If you have ACBD written down in front of you, you share your choice with 80 percent of the population. What you have done is arrange the photographs in reverse order of baby-faced-ness. Face A incorporates the stereotypical features of the mature face (smaller eyes; lower eyebrows; longer nose; "harder," more angular cheeks; and stronger chin) — and is thereby associated with dominance. Face D, on the other hand, incorporates the stereotypical features of the baby face (larger eyes, higher eyebrows,* shorter nose, "softer" cheeks, and smaller chin) — and is associated more with submission.

Actually, the way things turned out, all four cadets attained high military office. Below is how they looked at the peak of their careers, together with their identity:

Figure 2.10b — Portraits taken from the U.S. Army Military History Institute and the Center for Air Force History.

A: Lieutenant General Lincoln Faurer (Head of the U.S. National Security Agency).

B: General Wallace Hall Nutting (Commander-in-Chief, U.S. Readiness Command).

C: General John Adams Wickham Jr. (Chief-of-Staff, U.S. Army).

D: General Charles Alvin Gabriel (Chief-of-Staff, U.S. Air Force).

LIFE ON THE EDGE

In March 2004, Keith Lane's wife, Maggie, committed suicide by plunging to her death off Beachy Head — a high, sheer cliff on the south coast of Eng-

---

*Ever wondered why women "make up" their eyebrows in a higher position than they would otherwise appear naturally? Now you know.

land. They had been married for eight years. Beachy Head is a notorious spot for suicides. In 2004 alone there were some thirty recorded cases there. Earlier on the day of her death, Keith, a window cleaner from Eastbourne, had received a phone call from Maggie at work, but had noticed nothing unusual. Later he heard the news.

Several days on, after the initial horror had had time to sink in, Keith took a trip in his car. He felt drawn to the place where his wife had spent her last moments. Wanted to see with his own eyes what she had seen with hers. But after surveying the scene for a few gut-wrenching seconds, his attention was drawn to a woman. She was young, around twenty, with a pen and some paper in her hand. And was sitting on a bench, in a T-shirt, staring out to sea.

At first Keith thought nothing of it. She was a writer, perhaps. Or an artist. But then his mind started racing. What exactly *was* she writing, he wondered. Could she be another Maggie? Unable to settle, he decided to go over and talk to her. As soon as he reached her, he realized he'd made the right call.

His emotions still raw from the shock of Maggie's death, Keith was on dangerous ground. It had, after all, been only a matter of days. Yet despite the savage immediacy of his loss — in fact, maybe in hindsight *because* of it — he tried every trick in the book to talk the woman down. He even mentioned Maggie by name. But the more he pleaded, the more her resolve seemed to harden.

"My family couldn't give a damn about me," she told him. "Is there really any point in going on?" Eventually, she'd had enough. She shoved what she'd written between the wooden slats of the bench and started to make a run for it. Keith ran after her. The edge was no more than 15, maybe 20, meters away.

"All my schoolboy rugby training came back to me," he recalls, "as I dived for her legs and just hoped for the best."

That training came in handy. Keith, it turned out, succeeded in hanging on. Quite literally for dear life.

To say that the woman was grateful would, to put it mildly, be pushing it. Closer to the truth, she was livid. Several days later, when Keith went to visit her in the hospital, she slammed the door in his face. But eventually she got round to thanking him.

And then he had an idea. If he could save the life of one potential suicide, why not those of others? In fact, why not set up a watch on Beachy Head for exactly this reason? So he did.

In November 2009, some five and a half years after Maggie had ended her life, I spoke to Keith in Eastbourne. Life on the edge was over for him by then, and the Beachy Head Watch — six members strong and twenty-nine souls to the good — had been disbanded. Run-ins with the authorities had gradually taken their toll. As had the accusations.*

"What did you do," I asked him, "when you saw someone intent on taking their own life? What did you say to them?"

His answer was intriguing.

The best predictor of a successful talking down, he replied, was eye contact.

"When *I* looked at them and *they* looked at me," he told me, "that's when I knew I had them."

NOT A LOT OF PEOPLE KNOW THAT

Keith Lane's comments will come as no great surprise to anyone who has ever had to pull out into traffic at a busy road junction. The trick, as everyone knows, is to make eye contact with oncoming drivers. Once eye contact has been established, the chances of being let out increase dramatically. This is why it's much more difficult to enter a flow of traffic in sunny weather than it is when it's raining. Reason may dictate that motorists will be in a better mood when the sun is out, but nine times out of ten they'll also be wearing shades. Similarly, one fares better during the hours of daylight than at night. Put it another way. How many times have you accidentally blocked someone in and then gone to inordinate lengths to *avoid* making eye contact? See what I mean? Eye contact — just like cute looks — is a human key stimulus of persuasion.†

During the early days of his career, the British film actor Michael Caine

---

*Both the Coastguard and the Beachy Head Chaplaincy Team have allegedly accused Keith of not having the proper training to save lives, thereby placing his own life and, ironically, that of the person he's trying to save, in greater danger. Keith's response is pragmatic. "Every second counts," he says. "When you're in the business of saving lives, you often don't have the time to call for help. You have to *act.*"

†It's also an integral component of empathy — establishing rapport with others. One example of this can be found in military settings. Peacekeeping forces in Iraq, for instance, whose members wear sunglasses, report higher incidences of unrest, and incur more casualties, than those whose members keep their eyes visible.

had an intuitive grasp of the persuasive power of eyes. In a fiendish campaign aimed at raising his Hollywood profile, Caine began by training himself not to blink — to maximize the intensity of his closeups (when his eyes, magnified on screen, might be a couple of feet across) and reduce the chances of the director cutting away from him. An audience, Caine reasoned, enjoyed being paid attention to. And by actively endeavoring to fix them with his gaze, he could enhance the illusion that he actually found them attractive. Plus, of course, the opposite: how attractive *they* found *him*.

Empirical research substantiates Caine's chicanery. Take simple, everyday persuasion. Imagine I present you with an argument you don't agree with. I run you through the pros and cons, then attempt to get you on side. How can I improve my chances of eventually winning you over? One way, it's been demonstrated, is if I increase the amount of eye contact I have with you. Studies reveal that two people engaged in conversation don't look at each other in equal measure. The person who's listening looks directly at the person who's talking on average around 75 percent of the time — compared to just 40 percent eye contact from talker to listener. But up this latter figure to around 50 percent (any more and it starts to get uncomfortable) and a definite air of authority begins to filter through.

Statistics like these often come as a surprise to many people — though most of us, finding ourselves on the receiving end in such situations, certainly "get" it. Can such a small increment in eye contact really make a difference? The answer, almost always, is yes. Research has shown that eye contact can account for as much as 55 percent of information transmission in a given conversation — the rest being apportioned between "nonverbal auditory" (e.g., intonation) at 38 percent, and "formal" verbal content at just 7 percent. This is just one of the reasons why psychopaths — those undisputed kings of persuasion we shall be meeting later — enjoy the reputation that they do. On average, psychopaths tend to blink just that little bit less than the rest of us — a physiological aberration that often gives them their unnerving, hypnotic air.

"There is a road," G. K. Chesterton once said, "from the eye to the heart that does not go through the intellect."

THE EYES HAVE IT

Newborn babies have much in common with psychopaths. Ask any parent. They lack empathy, are superficially charming, possess not the slight-

est sense of the consequences of their actions, and are out purely for themselves. But they also share something else with their übercool, supersmooth
counterparts: the power to mesmerize with their eyes. This latter observation is well known to anyone who has ever caught the eye of an infant and
attempted to stare them out. Unless you're Uri Geller, forget it.

But babies don't just catch our eyes on a random basis. Studies have
revealed that such perceptual orientation is actually hardwired — in both
them *and* us. In 2007, a team from the University of Geneva compared the
degree of "attentional capture" by pictures of adult and infant faces on a
computerized reaction-time task. Results showed that reaction times were
slower on presentation of the infant faces — indicative of their greater "distraction" properties.

Conversely, psychologist Teresa Farroni of the University of London
showed paired photographs of faces to infants between the ages of two and
five days old. In one of the photographs the eyes were oriented forward, in
the other they were averted. What she found was remarkable: the babies
looked longer at the faces that they could make eye contact with than at
those they couldn't. A follow-up study also revealed enhanced electrical activity in the brains of four-month-olds on orienting toward faces with a direct gaze. And it would seem that we never really outgrow such a bias. Research conducted in art galleries shows that whenever we look at portraits
our attention is directed primarily at eye regions. But why? What are we
getting from this? Why the eyes as opposed to, say, the mouth or the nose?

One answer to this question has to do with survival.

There's nothing particularly special about eyes per se that attracts us, it's
more about where they're pointing. During the course of our evolutionary
history, the sudden orientation of eye-gaze to a particular location would
have acted as a powerful cue to potential sources of threat, and receptive-
ness to such cues would have conferred a considerable advantage when it
came to avoiding danger.

To demonstrate, Chris Friesen of North Dakota State University and
Alan Kingstone of the University of British Columbia have designed an experiment that captures precisely this power of attentional cueing. The procedure comprises three phases. In the first phase, schematically drawn faces
with featureless, blanked-out eyes appear at the center of a computer screen
for about half a second. In the second phase the pupils of the eyes material-
ize, orienting in one of three different directions: straight ahead, to the left,

or to the right (see Figure 2.11). Lastly, in phase three, a letter (an F or a T) appears on either the left- or the right-hand side of the screen — in other words, in either the same or the opposite direction to that in which the eyes are looking. What, Friesen and Kingstone wanted to know, would be the effect of these differently oriented eye regions on attention — specifically, on the way we process information in our environment? Would the direction of eye-gaze facilitate the speed at which individuals specified the position of the target? Or alternatively have little effect?

Figure 2.11 — Schematic faces with different gaze orientations similar to those used by Friesen and Kingstone.

The answer couldn't have been clearer. Performance speeded up. Results showed that individuals were faster at indicating the location of the target letter (left or right) on congruent trials (when the letter appeared in the *same* direction as that in which the eyes were looking) than on incongruent trials (when it appeared in the *opposite* direction to the cue). The eyes, ahem, "had it" — as the authors of the paper drolly pointed out.

HERE'S LOOKING AT YOU, KID

Friesen and Kingstone's cueing paradigm certainly offers a plausible explanation for our innate perceptual bias for eyes. But what does it tell us that we don't already know? Back in the 1960s, the social psychologist Stanley Milgram got a group of people together on a street corner. "Look up," he told them. What happened? Everyone else did the same.* And that's not all. Whether the cueing hypothesis tells us the whole story about eyes is also

---

*Milgram also found that the degree of conformity varied with the size of the group. When it was only the one person gazing skyward, the proportion of passersby who imitated them stood at 40 percent. This rose to 60 percent for three individuals, 75 percent for ten, and 80 percent for fifteen.

open to question. One thinks, for instance, of the profound attentional deficits found in those with autism.

Autistic infants prove the exception to the rule when it comes to focusing on the eye region of faces, attending instead to the area around the mouth. As they get older, autistic individuals also lack the ability to see, in both a cognitive and an emotional sense, where others are "coming from" — a deficit known as an absence of a Theory of Mind. Most children acquire the rudiments of a Theory of Mind by around the age of four, as assessed by a now classic experiment called the Sally Anne Task (Figure 2.12).

Figure 2.12 — The Sally Anne Task.

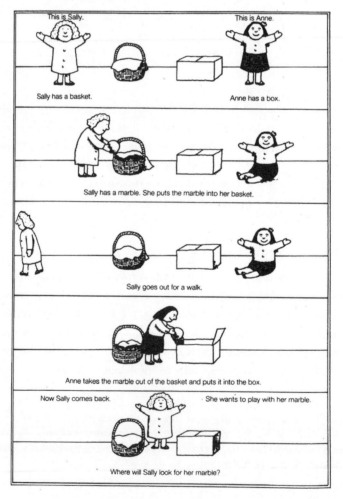

Up until the age of four, children will invariably give the wrong answer to the marble question: in the box. Because *they* happen to be familiar with the marble's new location it's inconceivable to them that others may not be. Eventually however, from about four onward, the correct answer gradually begins to emerge as the neurological rumblings of self-awareness proceed to disentangle our own mind from those of others.

Except, that is, in autism. From a clinical perspective, this is interesting. Disorders of the autistic spectrum are the only ones in DSM IV (the *Diagnostic and Statistical Manual of Mental Disorders,* published by the American Psychological Association) specifically characterized by an absence of a Theory of Mind. In addition, they are also the only disorders in which an inability to engage in eye contact presents as a key diagnostic feature. Might it be that our innate perceptual bias for eyes, as well as facilitating our propensity to detect threat, also foreshadows our capacity to "read" people? To infer the mental states of others?

Think, for a moment, of the potential long-term consequences that an inability to make eye contact might entail. If we lack the capacity to follow another's gaze, to glean even the most basic units of information about what that person may be looking at, how can we ever comprehend the notion that he or she may, in actual fact, possess a view different from our own? And if we cannot conceive of even these most fundamental gradations of autonomy, what hope do we have of ever fathoming the subjective — the hopes and fears, the intentions and motivations, of others?

EYE FOR INFLUENCE

Attentional cueing and the conveying of mental state are two of the most common explanations put forward for our attraction to eyes, and they cover, between them, a considerable amount of ground. Yet whether they cover all the ground is open to question. Why, for instance, does engaging in eye contact render persuasion more effective? And why are our eyes, with their oceans of white and tiny marooned irises, so radically different — in appearance, at least — from those found anywhere else in the animal kingdom?

The answers to these questions are rooted, I believe, in the state of total dependency in which we first enter the world. Newborns, we know, possess an innate perceptual bias for eyes. But could such a bias be a bit more complex than it seems? Might it not be for the eyes themselves — but rather for something, well, a little more fundamental perhaps? For the perceptual contrast between light and dark that characterizes their form? Could it be that what we've got going here is not, in actual fact, a unitary process at all but rather a two-tier model of influence? Where perceptual contrast engages the attention of the newborn, and where the newborn "locks on" with that vicelike grip of charm?

To take this second point first — the charm factor — one need look no further than the newborn itself. Not only are a newborn's eyes disproportionately large in relation to its face (the face, unlike the eyes, continues to grow after birth), but the pupils, too, are similarly disproportionate in relation to the sclera (the white, outer surface of the eyeball — see Figure 2.13).

Figure 2.13 — A baby's face in superstimulus format. Note the oversized eyes and the jumbo irises and pupils.

This latter observation is believed to reflect the relative inefficiency of the immature retina at capturing light. But research has also revealed that dilated pupils can serve a completely different function: forging the bonds of attraction.

"Which part of the human anatomy swells to twice its normal size when aroused?" the professor asks her class of first-year premeds.

There's an uncomfortable silence.

"Come on," she insists. "You must have some idea. Take a wild guess."

Still silence. Eventually, the sole male representative puts up his hand. Only for the professor to wave it back down.

"Forget it," she says. "It's the pupil!"

This is an anecdote that routinely does the rounds among the medical fraternity, and which is particularly well known, perhaps not surprisingly, among the female contingent. But I have a sneaking suspicion that were our

professor to have posed her question to the women of sixteenth-century Italy, she might've got more than she bargained for. In Italy at that time it was
customary for women to apply a few drops of the pigment belladonna — an
extract of deadly nightshade — to their eyes in order to widen their pupils
and make themselves more attractive to potential suitors. They certainly
knew what they were doing. But I doubt if they knew how it worked.

When shown pictures of two identical faces, one with dilated pupils and
the other without, and asked which of the two we rate as the more attractive, most of us will choose the face with the dilated pupils (see Figure 2.14).
Yet when asked to provide a specific reason for our choice, we just cannot
seem to come up with one. Intuitively, we feel that one face is "nicer" than
the other. More engaging, perhaps. Or friendly. But when it comes down to
it — well, it's anybody's guess.

Figure 2.14 — If forced to make a choice, which of these two faces do you find the more
attractive? Most people "think" it's the one on the right. But, when pressed, they are at a
complete loss as to explain why. Now take a look at the eyes.

In actual fact, the reason that we find faces with dilated pupils more attractive than those without comes down to reciprocity. Our pupils dilate
when aroused — when we encounter a stimulus that is either "easy on the
eye" or that we wish to learn more about. On occasions such as these we
strive, quite literally, to let as much of them "in" as we can. But not only do
such pupil responses occur automatically — outside of our conscious control — so does our receptivity to them in others. So, whenever we're presented with a picture of a face with dilated pupils we unconsciously infer
that the subject finds us attractive, and the law of reciprocity kicks in. We, in
turn, find ourselves more attracted to them.

This, incidentally, explains why we find dinner by candlelight a more romantic proposition than dinner in McDonald's. (It's one of the reasons anyway.) Under conditions of dim light, our pupils dilate to compensate for the reduced brightness of our surroundings, to allow more of what little light there is onto our retinas. So now you know (just in case you were wondering) why in many fast food outlets you practically need sunglasses to sit down and eat. It's because the accent is on *fast*. No lingering glances over the fries!

THERE IN BLACK AND WHITE

The eyes of the newborn seem custom-built to disarm. Their incongruous proportions and lagoons of latent empathy act as attention magnets — drawing us into their sparkling, innocent depths. But what of the flip side of the equation: an innate bias for contrast that allows them not just to meet our gaze, but to effortlessly lock on to it? Here, the evidence is similarly compelling. Studies have shown that when newborns are presented with two figures side by side — one depicting a dark circle within an oval (symbolic of an eye) and one a dark circle within a square — there's little to choose between them in terms of preference. However, when these two figures are presented, respectively, alongside an oval and a square on their own — minus the embedded dark circles — a somewhat different picture emerges. A strong preference obtains for the former, more "eye-catching" stimulus (see Figure 2.15).

Findings such as these seem to indicate that there's nothing particularly special about the eye itself that captures the attention of the newborn, but

Figure 2.15 — Infant preferences for shape and contrast combinations.

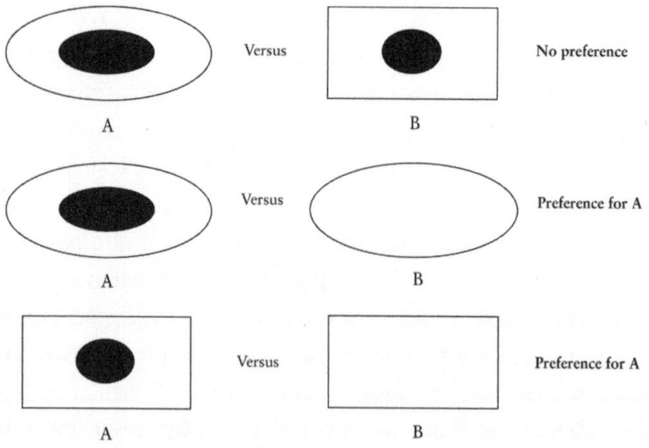

rather that it's the "novelty" of the stimulus — the perceptual contrast inherent to its appearance — that renders it unusually salient. And, moreover, that the greater this contrast is, the easier it is to delineate orientation.

"We have an uncanny ability," says the American ethologist R. D. Guthrie, "to determine the exact position of an individual's stare even though he is on the other side of the room — merely from judging the symmetrical alignment of a round pattern (iris) on a spherical one (eyeball). Exposure of the white sclera aids considerably in that ability. [The sclera] allow the transmission of fairly precise signals from the eyes." Which brings us, full circle, back to schematic faces and eye-gaze. And the adaptive benefits of ocularly cued attention.

So could a talent for split-second persuasion explain the newborn's soft spot for eyes, the same as it does their crying and their cutie-pie good looks? Do we have, in the newborn, persuasion in its purest form? A primordial ability to get the point across: "I am vulnerable. I am helpless. And you — yes, YOU! — need to do something about it"?

From the evidence put forward in this chapter, it could certainly be argued that way. Both the soundtrack and screenplay of neonatal behavior have been ingeniously choreographed, under the direction of natural selection, with one simple aim in mind: the immediate induction of nurturance and protection. Simplicity and empathy, so integral to the key stimuli of animal persuasion, are present in babies, too. The crying of a newborn, together with its looks, constitutes a primeval prototype, a prelinguistic paragon, of influence. And note, too, how perceptual incongruity plays its part: with looks, those wide, oversized eyes; with crying, those sudden, dramatic, unscripted shifts in pitch.

The moment we're born we're at our most vulnerable. Yet we're also, by exquisite evolutionary design, at the height of our persuasive powers.

Daryl, the South London mugger we met at the beginning of this chapter, certainly wouldn't argue. Were it not for an encounter with one of the world's greatest persuaders, the kind of bars *he'd* be used to by now wouldn't be dishing out mocha frappuccinos.

They'd be serving up porridge instead.

SUMMARY

In this chapter, we've continued our exploration of the biological basis of influence by looking at the extraordinary persuasive power of infants.

Newborn babies come into this world with just two simple aims — security and nurturance — and an overwhelming incentive to achieve them. Yet newborns travel light. Lacking the neural technology for sophisticated communication, they appear spectacularly underprepared for the challenge that lies ahead of them. How, without language, have they any hope of survival?

The answer, just as with animals, lies in key stimuli. An irresistible cry, a fundamental propensity to make eye contact, and an effortless cuteness, all converge into a psychological laser beam of influence: a beam that's trained directly on our brain's reward systems. There's not a schmoozer in history who could ever compete with an infant. We're never more persuasive than on our very first day on earth.

In the next chapter, we twiddle the persuasion spotlight in a slightly different direction. While sticking with the theme of immediate, incisive influence, we shift our attention to another kind of key stimulus — a kind that impacts not on ancient, subcortical reward systems but on cognitive process: the way our brains evaluate the world.

When it comes, as we've seen, to persuasion, animals and infants have two distinct advantages over the rest of us. First, they cannot think. Second, they cannot speak. But cognition and language have their own expressways of influence — and they're just as fast as any that went before.

We can, as we'll see, *learn* to be persuaded.

Question: Which type of crime are you *more* likely to be found guilty of if you're good-looking?

Answer: Crimes involving deception and fraud. It's precisely on account of the halo effect that good looks constitute one of the confidence trickster's most powerful weapons.

3

Mind Theft Auto

A man sets off on a fishing trip with his fishing rod in one hand and a suitcase in the other. Just as he's about to board his plane, he's stopped by one of the stewards. "How long is your fishing rod?" the steward asks. "Five feet," replies the man. "I'm sorry sir," says the steward, "but we can't allow anything longer than four feet on this flight. Can you fold it?" "No," says the man. "Then I'm afraid you'll have to leave the rod behind," says the steward. The man is furious. What good is a fishing trip without a fishing rod? he thinks. But, then, just as he's resigned to having to cancel the trip, he has an idea. A few minutes later both he, and his fishing rod, are safely aboard the flight. How does he solve the problem?

FISHER OF MEN

There's more to the brilliance of the streetwise, psychopathic hustler than just confidence, charm, and looks (though none of them do any harm). Don't believe me? Then meet Keith Barrett. For most of the 1980s and the early part of the 1990s, Barrett was a serial con man. He was shit hot at it, too. He was a master of the "long con" — elaborate and sophisticated stings usually, though not exclusively, confined to the corporate sector and involving large sums of money. Then one day, his number came up. One scam too many — city job, complex, worth a million, maybe more — necessitated he take an enforced sabbatical from his work. And when he came out five years later, having had an affair with the prison psychologist, he saw the world differently. He'd, ahem, found God.

Ever since he was at school, Barrett had always been good at getting people to do things. He regarded himself as a scientist, and the human mind his laboratory. And most of the formulas contained within psychology textbooks he'd managed to derive for himself. From first principles.

He was, you could say, a persuasion prodigy.

So it was little surprise that six months after joining his local church, the congregation had swelled to unprecedented levels and the ecstatic, if somewhat bemused, young minister was seriously considering locating to new premises. For which, thanks to Barrett, there were ample funds in the coffers. It wasn't so much a case of Barrett finding God, the minister enthused at the time. More of God finding him.

For Barrett, the reality was somewhat different. Far from the church constituting a new start, it merely presented a new window of opportunity. A new set of apparatus on which to try out the old experiments.

"Persuasion was, and still is, an addiction," he says. "I've got a cheating disorder. I get a high from getting people to do things that they otherwise might not. And the greater the amount of resistance I have to overcome to achieve that, the better it feels. Everyone shuts the door on Bible-bashers, right? So I thought to myself: I'm good at what I do. One of the best in the business. I've got a gift. A gift from God—who knows? It's just that in the past I've used that gift for my own ends. So why not do some good with it for once?"

He smiles.

"Or that's what I told the minister anyway. That pompous prick would have swallowed anything if it made him look good in front of his flock!"

Barrett's technique was unorthodox to say the least. It was also downright illegal. Jettisoning the sumptuous haute couture of the old days—the silk ties, the Gucci shoes, the Armani shirts, and the £2,000 Savile Row suits—he began by dressing down. In jeans, sneakers, and sport's shirt: the epitome of shabby chic. Such a retrograde costume change as this (made, he points out impishly, against all his *natural* sartorial instincts) flags up the extraordinary attention to detail, the predatory, arctic acumen of the ultimate persuasion virtuoso.

Here's Vic Sloan, another con man I spoke to, whose views on appropriate attire implicate color as well as style, and the hidden persuasive properties of a workaday pink shirt:

"The brain responds well to pink," he elucidates. "It's a scientific fact. Pink is a tranquilizing color. It produces a pattern of brain waves like no other. It stems from our evolution. Ancient man would have seen pink in the sky at sunset and sunrise — at which times, given the ambient light and circadian rhythms, it would have become associated with sleep and relaxation. So if you're trying to keep things on an even keel, pink is a good color to have around."

Sloan, in fact, might well be onto something here. The success of a particular hue of pink — Baker-Miller pink, or, as it's more commonly known, "drunk-tank" pink — in calming the mood of violent offenders has been scientifically documented in a number of U.S. studies. Reductions in anxiety levels, as well as in systolic and diastolic blood pressure, have all been reported in detainees held in rooms painted this color, in both civil and military detention centers.* In fact, following an experiment at the University of Iowa in which the locker rooms of visiting sports teams were daubed Baker-Miller pink so as to render their players less competitive, the Western Athletic Conference duly passed a law expressly prohibiting any further such forays into locker room interior design. The edict couldn't have been clearer. In future, it stated, the locker rooms of both the home and the visiting team could be painted any damn color they liked. So long as they were the same.

But I digress. Having sorted out his wardrobe, Barrett — suitably attired — would get down to business. And he would do so by deploying a technique that he describes as working the "Three A's" of social influence: attention, approach, and affiliation. Such a cocktail, according to Barrett, ships so much psychology into the brain's bloodstream that recipients lose all resistance to persuasion. It's the compliance equivalent of rohypnol. And it was all so easy.

Systematically targeting a preselected group of wealthy neighborhoods, Barrett — over a period of, say, six weeks — would gain covert entry to residents' cars. Then, once he'd turned on the sidelights, he would roll out the "good neighbor" trick of knocking on doors and informing them of their

---

*According to Alexander Schauss of the American Institute for Biosocial Research, dark brown or neutral gray flooring is best, and the optimal lighting level — "giving off a milder form of malillumination with color-rendering distortion peaking in the red-orange range" — around 100 watts. As to the science underlying this, speculation is currently ongoing — with research focusing on metabolic changes in neurotransmitters such as serotonin and norepinephrine, or in hormones serving the hypothalamus (the part of the brain that oversees the control of emotion). Pink, it would seem, is nature's Prozac.

"oversight." Having engaged his quarry in conversation (like most of his species, Barrett could sell shaving soap to the Taliban), he would contrive to explain why he just "happened to be passing," and request, perhaps, a small donation. Which, nine times out of ten, would be forthcoming. The request was perfectly timed so that it was made — nonchalantly — just when Barrett was walking *away* from his target. Further attention to detail.

"If they have to call you back, actually have to be *proactive* and ask you to stop on purpose," he explains, "without knowing it they've made a greater commitment than if you're just standing there passively waiting for them to give you something."

Some time later residents would notice the ad, for the church, which Barrett had persuaded the minister to place in the local paper. The laws of psychology then took care of the rest. The fact that they'd made a previous donation had educed in residents a token commitment to the church. And — what the hell — some of them actually went along to take a look at it. Not all of them, mind you. But some of them. More than would have done so had they just seen the ad and *not* made a donation.

And that, as they say, was that. Simple. Like taking candy from a baby. The church was packed to the rafters and Barrett had taken his cut.

The Good Lord works in mysterious ways, all right. And they don't, let me tell you, come any more mysterious than Keith Barrett.

THE RIGHT LINES

Keith Barrett is an evil, unfathomable genius. He's a psychopath. An evolutionary double agent. A predatory mind hacker who has made it his life's work to intercept and decode the psychological DNA of free will. The light switches of his brain are wired up in a different way from the rest of ours, and his neural meteorology is unpredictable. Yet Keith Barrett has something going on. He is, as well as being a ruthless, ice-cool hustler, one of the very best persuaders in the business. And what works for the psychopathic con man can also work for the rest of us.

I've been studying the principles of social influence for over fifteen years now. During that time I've come across a number of taxonomies that, like psychological string theory, purport to have reduced the science of persuasion to something you can slap on a T-shirt. Some of them, it has to be said, are better than others. But you want to know something? Barrett's Three

A's — attention, approach, and affiliation — are up there with the best of them, and provide the empirical backdrop to my own model of influence, which we'll encounter a little bit later.

"Look at it this way," says Barrett. "You know those cartoons of famous people you get in the newspaper? You can recognize who it is pretty much from nothing, the bare minimum of detail. Just a few key lines in the right place — but they *have* to be in the right place — and it's, like, 'Hey yeah, I got it!' Same with persuasion. You've just got to know where the brain's pressure points are. Where people's psychological blind spots are."

He is, of course, right about the cartoons. Take the example in Figure 3.1. We all know who it is, right? But just look at how much information is conveyed by so few elements. How an individual's entire physiognomy may be compressed into a few strategic squiggles. Exactly as Barrett said, it's not so much a case of how much detail you lay down. More how you lay it down.

Figure 3.1 — The economy of art: a few simple pen strokes speak volumes.

He's right about pressure points, too. And in the pages that follow we shall be looking at one or two of them. We shall decode the secrets of his shadowy street psychology. And, using his Three A's as a guide, get a con man's view of how the brain can be brought to its knees.

ATTENTION

During the course of any one moment, thousands of stimuli from the external environment come flooding into our brains. Yet we're only aware of — we only pay attention to — a handful. Consider what you are doing right now, for instance, reading this book. As your eyes move over the text, you're aware of the words and the pages on which they are printed, but probably not — until I mention it — the way the book feels in your hand. The reason for this is simple. The brain has a bureau that prioritizes information. And only information that's important to what we are doing — that's salient at the time — is allowed to filter through. The rest ends up in the shredder.

The fact that there are ways to hack into the brain's information bureau — and to rig what ends up in its in-tray — has been known since ancient times. In hypnosis, for example, the ability of the hypnotist to twiddle the knobs of consciousness, to orient it, like some neuropsychological satellite dish, is integral to the induction of trance. In magic, too, the attentional by-pass is common.

But cognitive distraction is also a part of persuasion. Here, as in magic, the power lies in misdirection — only linguistic rather than physical. The skilled persuader, just like the master conjuror, is also adept at controlling "where we look." And, more importantly, where we *think*. In fact, sometimes (remember Drayton Doherty and the lizard?) the line between magic and persuasion can be pretty tricky to draw.

A FOOL AND HIS MONEY

Three housemates walk into an electrical store intent on buying a cheap, secondhand television for their front room. They see one that they like and ask the shopkeeper how much it costs. The shopkeeper tells them that it costs £25 and they decide to split the cost equally. Each housemate hands over a £10 note and the shopkeeper goes to the back of the shop, where he keeps the cash register, to fetch them their change.

But as he does so he has an idea. Actually, he thinks to himself, if I tell them that I screwed up and the television costs £27, I can cream off some extra profit and no one will be any the wiser. And so this is what the shopkeeper does. He deposits the three £10 notes in the cash register and takes out five £1 coins, two of which he puts in his pocket. He then informs the three housemates that he's made a mistake on the price — the television costs £27, and not, as he'd originally told them, £25 — and hands each of them back £1.

The three housemates leave the shop quite satisfied — the television, after all, is still a hell of a bargain — and the shopkeeper is delighted with himself for having diddled them out of an extra couple of quid. Everyone's a winner.

But hang on a moment—there's a problem here surely? Let's run through it again.

The three housemates handed the shopkeeper £30 and the shopkeeper returned to them with five £1 coins from the till. Palming two for himself, and giving each of them back £1, means that each of the housemates has paid *how* much again for the television? Correct—£9.

$3 \times £9 = £27 + £2 = £29$.

Suddenly, we've got a pound missing.

So runs a notoriously tricky, yet fiendishly simple, problem. Many people—yes, myself included—have been floored by this dodgy psycho-arithmetic. But why? Why do so many of us slip up over the simplest of things? The answer to this question is sobering to say the least. The reason that we so often come unstuck over problems like this is because we have what we might call a "preparedness" to be deceived: a seriously impressive talent for being conned.

The way it works is like this. During the course of our evolutionary history, our brains, through the repeated assimilation of millions upon millions of microscopic slivers of information, have learned to take shortcuts. To use rules of thumb rather than work every single problem out from scratch. They have, to coin a phrase, "seen it all before." We make inferences about the world. Form expectations. Convert, to transpose LaPlace's famous observation, calculus to common sense. And on the basis of such expectations, we're vulnerable to sleight-of-mind.

"Life," said the writer Kurt Vonnegut, "happens too fast for you ever to think about it."

Natural selection agrees with him.

The efficacy of the missing pound problem is down to what Keith Barrett would call a "virus" of attention. Our brain is bamboozled into looking somewhere it shouldn't. Then—bam!—just like hypnosis, the unbelievable happens right under our nose. And there's plenty more where that came from.

Take, for example, the two photographs of Margaret Thatcher in Figure 3.2 (overleaf). OK, I know they're upside down. But apart from that, which one of them do you think offers the better likeness? The one on the left or the one on the right?

Figure 3.2 — The Thatcher illusion.

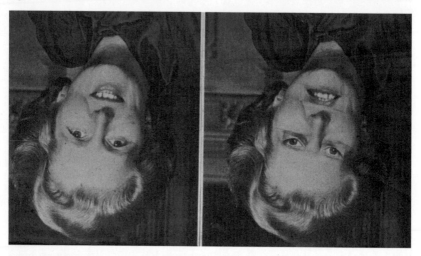

Now consider the following question: How many 9s are there between 1 and 100?

Go ahead, count them up — 9, 19, 29 . . .

Finally, read the following statement. Read it over once, at normal speed, and, as you are doing so, count the number of times the letter F appears.

Remember, read the statement only once.

FINAL FOLIOS SEEM TO RESULT FROM YEARS OF DUTIFUL STUDY OF TEXTS ALONG WITH YEARS OF SCIENTIFIC EXPERIENCE.

OK — how many times? Five? Six? Seven?

Actually, the correct answer is eight.

Don't worry if you got it wrong — you're in good company. A lot of people do. Even if you read it over again you're likely to screw it up. Most people, in fact, need at least three goes at it.

Same as with the 9s. How many did you get? Ten? Eleven, maybe? Perhaps if I told you that the correct answer was twenty you wouldn't believe me. Then again, what about 90, 91, 92, 93 . . . ?

Equally bizarre is the so-called Thatcher illusion. If you haven't done so already, try turning the Iron Lady the right way up.

And as for our fisherman at the beginning of the chapter, well, there's nothing here that the odd bit of Pythagoras can't sort out. It's called thinking *inside* the box:

Figure 3.3 — The 3-4-5 triangle.

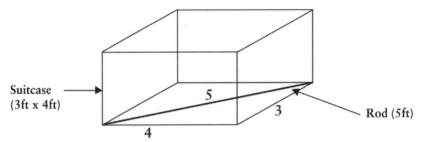

Suitcase
(3ft x 4ft)

5

3

Rod (5ft)

4

FOREGONE CONFUSIONS

Cognitive banana skins such as these, and our brain's unfortunate tendency to slip up on them, pertain in psychology to something we call "mental set." In everyday language mental set roughly translates as "frame of mind" and refers to the state of "autopilot" in which we, more often than we care to realize, go about our business.

In trems of the way in wchih we prcoess lagnuage, for istnance, scuh a sttae of atuopiolt revleas the paomnnehal pweor of the hmuan mnid. Aoccdring to rseearch at Cmabridge Uinervtisy, it deosn't mttaer in waht order the ltteers in a wrod are, the olny iprmoatnt tihng is that the frist and lsat ltteer be in the rghit pclae. The rset can be a taotl mses and you can sitll raed it wouthit porbelm.

Tihs is bcesuae the huamn mnid deos not raed ervey lteter by istlef, but the wrod as a wlohe. Tihs is epsceially the csae wiht retlviaely cmomon, soclaled "fuctnion" words lkie "of" — wchih maens taht one is lses lkiley, in scuh exmalpes, to prcoess the idnvidaiul cmpnonoets. Amzanig, huh?

Mental set also explains why psychological cat burglars like Keith Barrett can often persuade total strangers to do outrageous things for no good reason whatsoever. Ellen Langer, professor of psychology at Harvard University, once provided a classic demonstration of this in an experiment conducted, of all places, around a library photocopier. Armed with the knowledge that seemingly all of us detest both queue-barging and photocopying (alongside incest and murder, I wouldn't be at all surprised if anthropologists shortly announce the existence of two further universal taboos), Langer contrived two different varieties of excuse specifically designed to enable an associate to jump to the front of a photocopy line. The first of these excuses amounted to the fact that the individual in question was in a terrible hurry and only had one sheet to copy: standard, but effective, nonsense. The second constituted the following: *Please can I use the photocopier first because I need to use the photocopier?*

Incredibly, it turned out that this latter load of garbage proved to be just as effective as the first: definitive proof that reasons, under some circumstances, are processed in exactly the same way as fucntion words like "of." So long as they exist, we don't, in many cases, need to delve any further into the nature of their composition. They are present — peripherally so — and that's enough. They are, in short, an inviolable part of the syntax of everyday life. (Spot the typo? Go back four lines.)

BRIDE AND VROOM

Attentional viruses like mental set are something that all of us are susceptible to from time to time. And not just when we're waiting for the photocopier. As we saw a bit earlier with the Fs and the 9s, there are occasions in life when our brains get delusions of grandeur — when they make up their minds before *we* do!

Jim and Ellie Ritchie found out the hard way about mental set. Midway through their wedding reception at a posh hotel in Scotland, the best man suddenly discovered that the presents weren't where he'd left them. After discreetly making enquiries among the staff, a girl on reception finally cleared things up. An hour or so earlier there'd been a couple of guys and a truck. The guys had been wearing uniforms and had flashed a piece of paper. Don't worry, they'd said, it had all been arranged. Then they'd loaded up the truck and buggered off.

"Arranged?" asked the best man. "What do you mean, arranged?" The receptionist started to panic. "Arranged," she said. "The presents. They were all going back to the house." "What house?" asked the best man. "Er, I don't know," said the receptionist. "I thought . . . maybe . . ." She burst into tears. "The groom's?" Oh shit, thought the best man. The happy couple lived over 700 miles away.

Later, it all came out. The receptionist, it transpired, hadn't checked the men's ID. At the exact same time that the blaggers had sauntered in she'd been deep in conversation with the guy in 308. Some problem with room service, she said. She'd simply waved them through. Besides, she pointed out, why would she have suspected them? They looked the part. They acted the part. They were the part. Weren't they?

Sadly for Jim and Ellie — no, they weren't. And neither was the guy in 308. The room, it turned out, had been booked in the name of Smith. But at the time that Smith had been complaining about the bellboy, it was, you guessed it, empty. Smith had actually still to show. In fact, he never did.

Distraction crimes like this are bread and butter to your average con man. You don't even need to be good. When I told him about it, Keith Barrett laughed. This lot, he reckoned, would have been scouring the local papers — specifically on the lookout for weddings just like Jim and Ellie's. And then just taken their chances. Nothing personal, of course. Strictly business. You only need to look the part. Have a bit of confidence. And rely on the brain's unfailing capacity to jump to conclusions. All in a dishonest day's work.

(Want a demonstration of how — just like our receptionist — not paying attention makes us jump to conclusions? Have a go at the exercise at the end of the chapter on p. 97.)

But perhaps, on second thought, we shouldn't be giving the receptionist too hard a time. *She* didn't run off with the presents. In fact, if we're in the business of pointing fingers, the true architect of the heist is actually that shyster Smith — the elusive mystery guest in Room 308. Smith acts as an attention conductor: a concentration magnet who diverts all of the receptionist's available psychological resources away from the real problem and into a nonexistent one — in much the same way, for instance, that a bogus maintenance man might flash a fake ID, and then, immediately afterward, or better still simultaneously, initiate conversation to stop you looking too closely. Maybe your hair looks good. Or the car out front is cool. Next thing you know you're in debt. And the credit card people are calling. Words, especially nice ones, make excellent cognitive circuit breakers.

BRAIN STRAIN

The effects of distraction on our ability to make decisions illustrates our need to stay vigilant under pressure. To check that maintenance guy's ID. To verify his details. Cognitive resources are just like any other kind of resource. They are limited. And this, obviously, has implications for the way we allocate them. Ever wondered, when you read about them in newspapers or hear about them on TV, why many of the world's greatest con artists are also, more often than not, the world's greatest charmers? There's a reason for it. Basking in sweet talk doesn't come cheap. It's an expensive brain state to maintain, and constitutes a far greater *cognitive load*—places far higher demands on our limited cognitive funds — than does its opposite number: reality checking. Which means, in turn, that when we're lapping up the compliments we're eating up the assets. Brain assets. So there's less to invest in the business of critical thinking.

The principle of "cognitive load"—that the more operations our brain has to perform at any one time, the greater the drain on available resources — may be illustrated by a simple attention task, shown in Figure 3.4. First, cover up both figures with a blank sheet of paper. Then, when you have done so, uncover the figure on the left and locate the bold X within the display. Now uncover the figure on the right and do the same.

Figure 3.4 — X marks the spot.

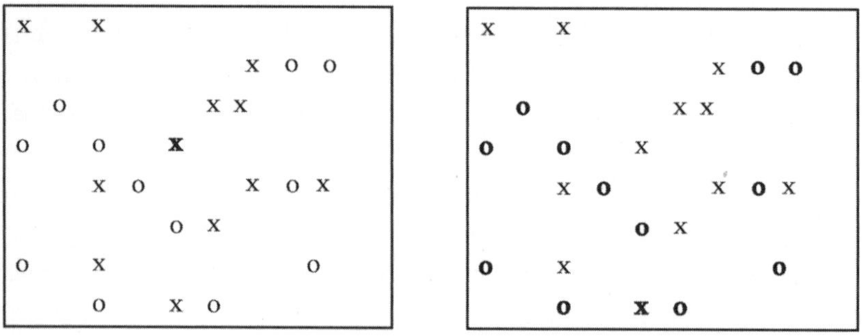

```
┌─────────────────────────┐   ┌─────────────────────────┐
│ X    X                  │   │ X    X                  │
│            X  O  O      │   │              X  O  O    │
│    O          X X       │   │    O            X X     │
│ O    O    X             │   │ O    O    X             │
│    X  O       X  O X    │   │    X  O        X  O X   │
│         O  X            │   │         O  X            │
│ O    X             O    │   │ O    X             O    │
│    O    X  O            │   │    O    X  O            │
└─────────────────────────┘   └─────────────────────────┘
```

Figure 3.4a　　　　　　　　　　　　Figure 3.4b

Did you find it easier to find the bold X the first or the second time round? I bet it was the first time. And the reason? In the second display, the demands on the brain's attentional resources are *double* what they are in the first. In the first, the brain only has to discriminate between contrast. In the second, between contrast and shape.

Yet cognitive load can also work in our favor. Ironically (given that it's a favorite device of con artists), putting the brain under a bit of pressure — increasing the amount of work that it has to do — can also be useful if we want to turn the tables: find out if someone is lying. Which, when you think about it, actually makes perfect sense. The more you give someone to mull over, the fewer resources his or her brain has available to help conceal the truth. In fact, in police interviews and military interrogation, this is standard practice. Tried and tested techniques such as the exchange of "knowing" looks between officers, close physical proximity, robust handling, positioning of detainees away from so-called control triggers such as light switches and door handles, officers being called out of the room on the pretext of "new information" coming to light, "incriminating evidence" (sometimes just blank sheets) being placed facedown on the table, and dossiers with everything but the suspect's name obscured are all, on timely introduction, powerful conduits of "diversionary" persuasion.

Just so you know. For next time.

APPROACH

No two people see exactly alike. That's what the old empiricists used to tell us, and they're right. Low-level differences in perceptual awareness do exist between individuals. On the other hand, however, when it comes to the way we see the world in general, we have a lot more in common than we think.

Consider, for example, the following. Imagine that someone hands in a lottery ticket bearing the numbers 1, 2, 3, 4, 5, and 6. Which of these two scenarios would cause you greater amusement:

If the winning sequence turned out to be 4, 14, 22, 33, 40, and 45?

Or if it was 7, 8, 9, 10, 11, and 12?

Nearly everyone says the latter (unless it was his ticket). But why? In actual fact, the probability of the winning ticket bearing *either* of these sequences is the same.

Take another example. Imagine you're at a raffle and the winning number is 672. When would you feel harder done by? If you had 671? Or, alternatively, 389?

These two instances reveal something rather interesting about our brains. They are slothful creatures of habit. Rather than preparing decisions from scratch using fresh, seasonal ingredients, they prefer the readymade variety — chockfull of conjecture, assumption, and prepackaged, freeze-dried reasoning.

Such information, in the wrong hands, can be dangerous. In sport, reading one's opponents, knowing what makes them tick, knowing how they're likely to approach a given play, is the goal of every competitor.

It's exactly the same in persuasion.

TALL STORY

Imagine that you're working for a marketing company and you're processing a series of surveys completed by a random sample of adult men in the United States. One of the respondents, who has put his height down as over six feet five inches, has been less than precise when it comes to the question about employment. It's not clear whether he has put a cross by "bank manager" or by the item next on the list: "basketball player." It's up to you to make the decision for him. Which profession are you going to go for?

If you chose "basketball player" — congratulations! You share your answer with 78 percent of first-year undergraduates at the University of Cambridge. Unfortunately, however — just like them — you're wrong.

Let me start by asking you this. In the general population of the United States, who do you think there are more of — professional basketball players or bank managers? The answer, I think you'll agree, is bank managers. Let's put an arbitrary figure on each. Let's say there are 300 basketball pros and 15,000 bank managers.

OK, now out of those 300 basketball pros, how many do you think are over six feet five? 60 percent? 70 percent? Let's say 70. Which means, by my calculations, that there are 210 professional basketball players in the United States who are over six feet five inches.

Now let's take the bank managers. Out of those 15,000, how many of them do you think are over six feet five? On this occasion, let's go for a conservative estimate and say 2 percent.

Yet even if only 2 percent of bank managers are taller than six feet five, that still accounts for 300 people — which means that there are 90 more bank managers over six feet five inches than there are basketball players.

Oh, dear.

What's just happened here with the basketball player and the bank manager introduces us to the second of Keith Barrett's key ingredients of influence — *approach*. Approach, in Barrett's system, refers to our attitudes and beliefs about the world. Or, more specifically, how these attitudes and beliefs impact on the kinds of decisions we make. The reason that we perform so poorly on tasks such as this is actually quite simple. It has to do with the way our brain processes information about the world. The way it files its paperwork.

In the example above, the brain is required to solve a mystery. Its detective skills are called upon. The "crime" is being over six feet five inches tall, and there are two "suspects" — the bank manager and the basketball player. In light of this initial information, the brain runs a preliminary check on its database — "Just routine, sir" — and in so doing, something rather interesting pops up on screen. The basketball player has a number of "previous convictions" for exactly such a crime. There is, in contrast, no record of the bank manager whatsoever. So, faced with such "evidence" as this, what does the brain do? Well, like any seasoned detective, it hauls the basketball player in for questioning and decides not to bother with the bank manager.

The analogy of the brain as forensic database is not one that you're likely to come across all that often in the pages of psychology texts. And there are probably, somewhere, some very good reasons for that. But for present purposes, it fits rather well. For, just like such a system, our brain profiles incoming information according to perceived probabilities and known associates. It engages in speed reasoning. Or, to use the proper terminology, it employs heuristics.

In the category of height, for instance, six feet five inches or over is a "known associate" of the basketball player. As such, it seems far more likely that the two of *them* will be working together than it does that six feet five and the bank manager will be in partnership. In the more orthodox language of cognitive psychology, we form, on the basis of previous experience, a *schema* or an *associative network* of basketball players and bank managers — a general concept of "who they are" — and these schemas are underwritten by certain ultrasalient descriptives such as "tall" or "wears a collar and tie." Once such exemplars are entered into the system, those on file who "match the description" are flagged up as deserving of closer scrutiny. Yet sometimes, as we've just seen, the real culprits slip through the net.

MAGIC SQUARE

Heuristics are pretty much indispensable during the course of everyday life. They're the cortical equivalent of the fixed action patterns that we saw in animals in Chapter 1 — short, sharp bursts of automatic behavior, triggered by the presence of a key stimulus. They provide smooth, underground slip-roads through the middle of rush-hour consciousness that enable the brain to shoot across town in a hurry. But these slip-roads are dangerous. And con men like Keith Barrett know them like the back of their hands. They are fast. They are dark. And they are covered in psychological black ice. And accidents, as we've just seen, are commonplace.

"The brain," says Barrett, "is like a Snakes and Ladders board. You can go the long way round and pass through every number. Or you can hit the ladder on nine that takes you up to ninety."

It's a ladder he's been up a lot.

From an evolutionary perspective, there's no way around any of this. With the possible exception of some of the new self-serve check-in con-traptions they've installed at Heathrow airport, the human brain is the most complex piece of machinery we are ever likely to see. Yet we're still hard-wired to be fall guys. Bizarrely, taking everything into consideration, there is method in this madness. Irrespective of how brilliant our brains may hap-pen to be, we cannot double-check every single thing that occurs to us on the basis of veracity. Life is, quite literally, too short. Instead, rather like a physician diagnosing an illness (or a detective solving a crime), we must rely on "tried and tested" presenting symptoms — on superinformative key stimuli — to guide our behavior. Not spots like the Herring Gull. Or quonks like the Louisiana Bell Frogs. But rather the deep, accumulated wisdom of learned associations. And sometimes, because of our brain's unfortunate penchant for shortcuts, we get the answer wrong.

GRAPE EXPECTATIONS

What happened earlier with the bank manager and basketball player is known in psychology as the *representativeness heuristic:* a rule of thumb by which our brains make inferences about the probability of a hypothesis by considering its fit with data already available. And it doesn't just happen when we're sitting down filling out forms.

In a study which looked at the effects of expectation on taste, Hilke Plassman and her colleagues at the California Institute of Technology sneakily switched the price tags on a middling bottle of cabernet. Some vol-unteers were told that the bottle was worth $10. Others were told it was $90. Did this difference in price make a difference to how the wine tasted?

You bet it did.

Participants who were told that the bottle was worth $90 pronounced it a far better wine than those who thought it was worth $10.

And that wasn't all. Subsequently, under fMRI conditions, Plassman found that this simple sleight-of-mind was actually reflected anatomically — in neural activity deep within the brain. Not only did the "cheaper" wine taste cheaper and the "dearer" one, well, dearer — the latter also excited the medial orbitofrontal cortex, the part of the brain that responds to pleasurable experiences.

Similar results have also been found with experts. Cognitive psychologist Frederic Brochet of the General Oenology Laboratory in France took a midrange Bordeaux and served it in two different bottles. One was a kick-ass grand-cru. The other a *vin de table*.

Would the bottles have an impact on the connoisseurs' rarefied palates? Or would the wine buffs smell a rat?

Not a chance.

Despite the fact, just as in the Plassman study, that they were actually being served the same vintage, the experts appraised the different bottles ... differently. The grand-cru was described as "agreeable, woody, complex, balanced, and rounded" — while the *vin de table* was evaluated less salubriously: "weak, short, light, flat, and faulty."

John Darley and Paul Gross at Princeton University have taken things one stage further and demonstrated this effect in a study on social class. In their take on the paradigm, participants evaluated the performance of a child as she worked on a series of math problems. The participants were divided into two groups. One group was told that the child was of low socioeconomic status (SES), while the other was told the opposite: that she was of high SES.

Who do you think rated the child as more intelligent? Correct — those who were told that she was of high SES. Moreover, this simple belief bias went way beyond just math — it was sufficient to account for judgments of *general* intelligence. Those who believed the child to be of low SES rated her performance as below average, while those who believed her to be of high SES rated it as above. Socioeconomic status — in wine, in people, in anything — constitutes a key stimulus of approach. And colors our perception more than we care to realize.*

---

*In Appendix 1, you can test this out for yourself by giving the brief character sketches — and subsequent impression-formation task — to your friends. You'll be amazed by what something as ostensibly simple as the kind of house we live in says about us to others!

SOMETHING IN MIND

Expectations, of course, don't just have an impact on *perceptions* of performance. They can affect performance itself. Take academia, for example. On sitting for the Graduate Record Examination (GRE) in the States, black volunteers perform significantly worse if they're told, prior to testing, that the exams are indicative of a person's level of intelligence. Findings such as these, where notions of inferiority concerning a *group* to which we belong can significantly affect our ability as *individuals,* reflect what is known as "stereotype threat" — "stereotype lift," in contrast, describes the opposite: when a sense of in-group superiority actually *facilitates* performance.

Margaret Shih at Harvard has demonstrated this empirically. In a study which looked at women from Asian backgrounds, Shih found that when the women were primed to think of themselves as "women" they performed *worse* on math tests than men — thereby confirming the familiar "male/female brain" stereotype. Conversely, however, when they thought of themselves as "Asian" they actually performed *better* than men — Asians, stereotypically, enjoying generally higher math cred than other ethnic groups. Jeff Stone, at the University of Arizona, reports similar findings in sport. When golf is presented as a test of athletic ability, black golfers outperform whites. But when the game is depicted as a showcase of cognitive strategy, the trend mysteriously reverses: whites do better than blacks. Race, like socioeconomic status, is another key stimulus of approach.

Related to the concept of representativeness is that of availability. If representativeness refers to the way our brains make *probabilistic* inferences about the relationships between variables (e.g., occupation and height; socioeconomic status and academic ability), availability describes a more "temporal" kind of inference: our tendency to confuse how *frequently* an event occurs with the ease with which examples of it may be brought to mind.

To illustrate, consider the following cheery pronouncements:

More people die from firearms than from asthma.
More people die from cancer than from stroke.
More people die in accidents than from emphysema.
More people die from homicides than from floods.

How many of these estimates do you agree with? Might it be all of them by any chance? If it is then you're in good company. Most people think the same. But actually, you're in for a surprise. All of these estimates are wrong. Some of them by miles. Now ask yourself this. Of the kinds of fatality just described, which do you hear most about? Which are the most "available" in your memory?

It's difficult to convey the power of the availability heuristic without a concrete example. So let's take a look at one right now. Below is a list of names. Read them over carefully, and then as soon as you've done so, cover them up with a sheet of paper:

Elizabeth Taylor	Mark Radcliffe	Michelle Obama	Hillary Clinton
Andrew Marr	Raymond Carver	Agatha Christie	Stuart Rose
Angelina Jolie	Madonna	Norman Foster	Amy Winehouse
Ian Poulter	Margaret Thatcher	Cheryl Cole	Chris Martin
Oprah Winfrey	Anthony Eden	Steve Jobs	Paul Simon
Robert Frost	Kate Moss	Rowan Williams	Britney Spears
James Nesbitt	Barbra Streisand	Damien Hirst	Bruce Chatwin
Ruby Wax	Florence Nightingale	Ranulph Fiennes	Princess Diana

OK. Now that you've read the names try to recall as many of them as you can. Then estimate whether there were more women on the list, or men.

Only when you've made your estimate should you read on . . .

Made your estimate? Great — what was it? More women than men by any chance? That's fine, it's what most people say. But now take another look at the list. Count up the names. Funny, huh? The sex ratio is exactly the same. There are precisely as many men on the list as there are women. But notice anything else — how the women are more famous, perhaps?

Here's another example. Give yourself sixty seconds to come up with as many words as you can that conform to the pattern $-----n-$. When you've finished, repeat the test only this time with the pattern $----ing$.

Chances are you came up with more words the second time round than the first time. But actually, you *shouldn't* have. Take a closer look and you'll notice that the first example is, in fact, identical to the second. Only with the letters "i" and "g" blanked out. Which means that any word that conforms to the second example must automatically conform to the first. Which means — that's right — that words fitting the first template are actually more common.

But words that fit the second template more easily spring to mind.

SOFT TOUCH

Our hardwired propensity to jump to conclusions, to respond, completely instinctively, to what we might refer to as "conceptual" key stimuli — constructs containing high levels of representativeness and availability, for instance — provides easy pickings for the sharks of social influence. And, of course, for the rest of us. As Keith Barrett somewhat chillingly points out, if you know where the ladders are and can handle a die, the game — and for these guys it *is* a game — is over pretty quickly.

Take Shaffiq Khan, for example. Khan, like Barrett, is another supersmooth psychopath. But Khan, unlike Barrett, concentrates his killer persuasion predominantly on individuals rather than on large corporations. Khan's motivation is the high life. "There's no level of luxury to which I cannot aspire," he tells me over lunch in a chic London restaurant. From the look of him — Rolex, Porsche, Armani — it's difficult to disagree.

Khan's modus operandi is disarmingly simple. He crisscrosses the globe as a glamorous entrepreneur (which, in a way, I guess he is), exhibiting a ruthless dedication to the art of looking good. He stays in the smartest hotels. Frequents the hippest bars. And he always flies first class. It's here, in the world's most exclusive hangouts, that he plies his deadly trade: charming, and then seducing, sometimes female staff members, sometimes fellow clients, whom he subsequently cleans out.

Khan is coy about his seduction techniques. But he offers the following insight:

> Touch is important — physical contact. You see it in primates as they groom each other. It's a method of ingratiation. You scratch my back, I'll scratch yours. Now with humans, ingratiation is usually something that people of lower status do to people higher than them.* Again, exactly like in primates. It's hardwired into us from evolution. They'll try

---

*Just like Vic Sloan with his pink shirts, Khan may well be on to something here. In his book *The Right Touch: Understanding and Using the Language of Physical Contact* (Cresskill, NJ: Hampton Press, 1994), Stanley E. Jones describes an experiment conducted in a public health organization. He writes as follows: "The group studied was a detoxification clinic, a place where alcoholism is treated. This was an ideal setting in which to study status, sex roles, and touching. . . . [The] findings showed two clear trends. First, women on the average initiated more touches to men than vice versa. Second, *touching tended to flow upwards, not downwards, in the hierarchy*" [author's emphasis].

to build bridges, store up favors by being tactile. So our brains are pro-grammed to expect lower-status people to be more touchy-feely. Re-verse that expectation — which is what I do: I always initiate contact with a gentle touch of the arm or on the small of the back — and it's very powerful. It says: *you* are of value to *me,* rather than the other way round. And they think — why would *I* be of value to *him*? He's already got everything he needs. He must really like me.

Though on the surface mundane, what Khan does with his hands is actually quite intoxicating. For brownnosers, both representativeness and availability heuristics involve people of lower status sucking up to those of higher status. But confound that expectation, chuck in a bit of antith-esis — switch on the hazard signs on those superfast cognitive express-ways — and suddenly, dramatically, we have to slam on the brakes. We have to make sense of the snarl-up.

Psychologist David Strohmetz and his colleagues at Monmouth Uni-versity have demonstrated a principle very similar to the one Khan uses. Only in Strohmetz's case, the purpose isn't to fleece people but rather to in-crease tips in a restaurant. Strohmetz divided diners into three groups de-pending on how many sweets each was given at the end of their meal. To one group of diners the waiter gave one sweet. To another, he gave two. And to the third — and here's the deal — he did this. First he gave one sweet and then walked away. Then he turned back around (as if having second thoughts) and added another. So one group got one sweet. And two groups got two. But the two who got two got them in different ways. Got it?

Did, as Strohmetz predicted, the number of sweets and the manner in which they were given have any bearing on tip size?

They sure did. Compared to a control group of diners who got *no* sweets at all (great!), those who got *one,* tipped, on average, 3.3 percent higher. Not a bad investment for an outlay of less than a dime. Similarly, those who re-ceived *two* sweets left, again on average, an extra 14.1 percent on the table. Even better. But the highest increment of all was shown by those who re-ceived first one sweet, then another — a staggering hike in generosity of 23 percent!

That unexpected, and seemingly inexplicable, change of heart (Hey, for you guys here's two instead of one!) cut through those purse strings like a hot knife through butter — in exactly the same way that Shaffiq Khan's un-expected, and seemingly inexplicable, use of touch cut through the purse strings of his unsuspecting victims.

Evolution, on the one hand, has programmed our brains with a fast lane, with cognitive heuristics like representativeness and availability. But it has also equipped them with another, more specialist kind of program: an inbuilt facility to make sense of the world — to convert data into mean-ing, and the chance and the random into pattern. Play one of these pro-grams off against the other — by violating expectation — and the system, momentarily, goes down. Dangerous times if you're dealing with someone like Khan.

AFFILIATION

A king once paid a visit to his country's prison, and listened intently as one inmate after another begged to be released on the grounds that he was inno-cent. Suddenly, the king noticed a withdrawn and dejected prisoner sitting by himself in the corner.

The king approached the man, and asked him: "Why do you look so troubled?"

"Because I'm a criminal," the man replied.

"Is that right?" asked the king.

"Yes," said the man. "That's the truth."

Impressed by the man's honesty, the king ordered his release with the following observation: "I don't want this criminal to be in the company of all these innocent men. He would be a bad influence on them."

"No man is an island," wrote the poet John Donne who, with apologies to Kurt Lewin, should really be hailed as the father of social psychology. Our behavior, since ancient times, has been inextricably interwoven with the behavior of those around us. And the greatest influence on any of us is others.

GROUP INFLUENCE

We humans are hardwired to stick together. To form groups. Not only that, we're also hardwired to favor the groups we belong to over those we don't.

For no apparent reason.

That may seem an odd thing to say. But it's true. Back in the days of our ancestors, membership of a group constituted the very first life insurance policy. And, boy, did we need one. We've been renewing that ancient premium ever since.

In 1971, the late Henri Tajfel of the University of Bristol conducted an experiment which illustrated precisely the kind of arrangement that we have with natural selection. Indeed, the experiment proved so revealing that it's since become a classic, lending its name to an entire paradigm within social psychology: the *minimal group* paradigm.

What Tajfel did was this. First, he took a sample of secondary-school students and showed them a display of dots.

"How many dots do you see on the screen in front of you?" he asked each one of them individually.

Because there were quite a few of them — dots, that is — and the time allowed was less than half a second, the students had no idea as to the accuracy of their estimates. But they provided them anyway — and this deliberate manipulation conveniently enabled Tajfel to divide the students up into groups. Two, completely arbitrary, "minimal groups": underestimators and overestimators. "Minimal" because the categorization had been trumped up on the basis of a trivial and nonexistent difference. "Group" because there were a number of them.

Once the categorization had been made, Tajfel then asked each student to allocate points — which they'd been told equated to money — to two of their fellow participants in the study. These other students were solely identifiable by code — they were anonymous, in other words — and by one or other of the following labels: "of your group" or "of the other group."

Would the simple fact that the students were members of one group as opposed to another bias their allocation of points? The answer — which Tajfel and his colleagues correctly predicted — was yes. And by quite some margin.

Prospects of financial reward were positively showered on the members of one's own group, and trickled — if that — toward those of the other. What was more, this was in spite of the fact that, prior to showing up for the study, none of the participants had met each other before. And, in addition, would probably never have done so again. Ergo, no you scratch my back, I'll scratch yours. Rewards, pure and simple, had been doled out on the basis of label: my lot as opposed to yours.

THE WRONG LINES

It's not difficult, when one looks at the world, to appreciate the power of the in-group bias. One only has to turn up at a football match to understand that much. But what may come as a surprise is that the effects of group membership go far beyond a superficial favoritism. They extend, in fact, to how we actually *see* things.

Take the following simple problem illustrated in Figure 3.5. Which of the three perpendicular lines shown in Box A is the same length as that in Box B?

Figure 3.5 — A simple perceptual judgment task.

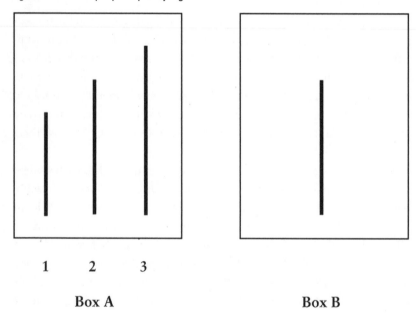

Piece of cake, right? The middle one — line 2. You'd need your eyes tested if you got it wrong. Yet you know what? I bet I could persuade you to do exactly that — get it wrong — by using a very simple maneuver. Don't believe me? Luckily, you don't have to.

In 1955, the American social psychologist Solomon Asch performed exactly such a feat in one of the classic early demonstrations of the power of conformity. What Asch did was as follows. First, he got a group of nine people together in front of a slide projector. Secondly, he presented the group with a series of eighteen line-judgment tasks identical to the one just shown in Figure 3.5. Thirdly, prior to the start, he instructed eight of the nine group members (his evil coconspirators) to give the same predetermined wrong answer on six of the eighteen comparisons. Lastly, he sat back and watched what the ninth guy would say. Would he, in the face of such unanimous disagreement, stick to his guns and give the correct — and obvious — answer? The one that was, quite literally, staring him in the face? Or would he cave in to peer pressure and go against the evidence of his senses?

What Asch found was astonishing. Of those who participated in the study, 76 percent gave at least one wrong answer during the course of the proceedings. Think about that for a moment. They gave the wrong answer — over three-quarters of them — to a task as simple as the one just demonstrated. The conclusion was as clear-cut as it was scary. So great is our desire to fit in that most of us are prepared to disbelieve even what our own eyes are telling us, so as not to stand out. Majority opinion is one of the most powerful forces in the universe. Few of us, it would seem, have the psychology to hold out against it.*

Why do you think that canned laughter is so popular on TV sitcoms? Or that, on the campaign trail, not all of the applause is — shall we say — as spontaneous as it seems? Devices such as these sneak into our brains through the back door and engage our emotions in neurological pantomime. They persuade us (or rather, help us persuade ourselves) that whatever or whoever we're watching is more humorous, more entertaining, or more interesting than they really are. I mean, if everybody else is laughing or applauding or jeering — then why not us?

But there's more to this pantomime than meets the eye. The affirmation, or derogation, of a communicator by an audience molds our perception

---

*A recent study by Vasily Klucharev at the Donders Institute for Brain, Cognition, and Behavior at Radboud University in the Netherlands has revealed a possible neural correlate of conformity. In a facial attractiveness judgment task, Klucharev and his coworkers have found that individual conflict with group opinion triggers increased activity in both the rostral section of the anterior cingulate cortex, and the ventral striatum: areas of the brain implicated in error detection and decision making under unusual circumstances. (See the article by Vasily Klucharev, Kaisa Hytönen, Mark Rijpkema, Ale Smidts, and Guillén Fernández, "Reinforcement Learning Signal Predicts Social Conformity," Neuron 61 (1) (2009): 140–151.)

not just of how funny or entertaining he is, but also how influential. How suitable she may be for office. And it's then that things start to get serious.

There are few better examples of influence such as this than a study conducted in 1993 during the third Bush-Clinton presidential debate. Three groups of thirty students each were carefully assembled on the basis of political allegiance. The first group (which actually consisted of only twenty "real" participants—a mixture of Republicans and Democrats) concealed ten plants who cheered Bush and jeered Clinton. The second group—again consisting of only twenty "real" participants—concealed ten plants who, you got it, cheered Clinton and jeered Bush. A third group—the control—remained neutral. What impact, respectively, would the two lobbying factions have on "real" participants' ratings of the candidates?

The results from the pro-Clinton lobby group (the group with the ten plants who cheered Clinton and jeered Bush) are shown in Figure 3.6. They proved a real eye-opener.

Figure 3.6

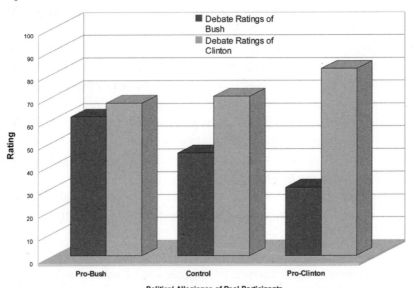

On the right—among the *real* Clinton supporters—we observe Clinton's ratings soar when Bush is jeered and Clinton cheered. No surprises there. But just look what happens on the left—this time in the Bush

camp — under the same conditions. Incredibly, even "real-life" Bush supporters evaluated Clinton more favorably than their own man when the
former was booed and the latter applauded! Sometimes, it would seem, the
way we see others depends on nothing more substantial than the way *others*
see others.

POUND FOR POUND

Examples such as this are well known within the science of influence as instances of social proof. Social proof is what Keith Barrett might describe
as a virus of affiliation — and occurs in ambiguous social situations when
one is unable to determine what, precisely, is the "done thing." We've all
been there. I guess the classic example is sitting down to one of those fussy,
thirty-five-course dinners only to discover that the display of cutlery bears
a frightening resemblance to something we might find in an operating theater. What do we do? Where do we start? Which knife do we use for the butter? What's that funny pointy-shaped thing next to that other funny pointy-
shaped thing with the little hook on the end? Making the assumption that
the guy sitting next to us has seen it all before, most of us resort to guerrilla
tactics. We scrutinize him out of the corner of our eye. Observe where his
fingers alight upon the silverware . . . and then apprehend that very same
implement ourselves — totally unaware that our worldly bon vivant has just
spent the past five minutes secretly scoping *us* out.

A fascinating demonstration of the power of social proof occurred not
so long ago on American TV. Colleen Szot, an "infomercial" writer, completely obliterated a home shopping channel sales record by changing just
three words of a now familiar sales pitch. The record had stood for almost
twenty years. Sure, the channel featured all the usual marketing parapher-
nalia: celebrity endorsements, catchy sound bites, and an audience that
looked like it was on speed. But, remarkably, it wasn't any of these that finally sent sales through the roof. Instead, it was a change for the *worse*. Or
so, on the face of it, it seemed.

Szot's masterstroke was to tweak the standard call-to-action line "Operators are waiting, please call now" to "If operators are busy, please call
again." At first glance, such nuancing appears disastrous. How does alert-
ing customers to the prospect of inconvenience — dialing and redialing the
same damn number — possibly translate into increased sales? But logic, in

this case, leaves much to be desired — and fails to account for the magic of social proof.

Think about it. What image springs to mind when you hear the phrase "Operators are waiting, please call now"? Legions of bored telephonists staring into space? If so, then despite all the flashy, trashy merchandising, your impression of the product is negative. It's an impression of low demand and poor sales. Why the hell would *you* want to buy it if nobody else does?

Now ask yourself this. What comes to mind when you hear the phrase "If operators are busy, please call again"? A buzzing call center full of overstretched staff struggling to keep pace with demand? Now that's more like it! If everyone else is getting in on the action — then *you're* sure as hell not going to miss out!

Exactly the same principle works on eBay. Analysis of online auctions reveals something primal, profound, and fundamentally potty about consumer behavior: if you want to flog that Rembrandt you found in the attic, start at $10! The psychology here is actually quite straightforward. Kicking things off with a low opening bid attracts a greater number of people to the auction — which, in turn, makes the product appear more desirable. This generates even more bidders who, with every mounting offer, increase not only their financial investment in the product but their emotional investment, too.

A friend of mine who teaches decision science demonstrates this in the lecture hall. Not, of course, with a Rembrandt, but with a £1 coin. Every year at the beginning of the winter term, he stands in front of a packed auditorium of freshers and announces that he's going to auction . . . £1. What am I bid? he asks. The auction has two simple rules. The first rule — like that of any auction — is that whoever puts in the highest bid gets to keep the merchandise. No problems there. The second rule is that the person who enters the next highest bid forfeits that bid to the auctioneer. No problem there either — so long as you win.

Ever since the inaugural £1 auction several years ago, my friend's students, without exception, have remained neuroeconomically impervious to the juxtaposition of these two conditions. They see it every time as a gilt-edged opportunity to get something for nothing. Or, if not for nothing, then at least for less than £1. He must be crazy, they think. The first bid usually weighs in at a penny. Surprise, surprise. Then two pence. And three pence. And so on. Everyone's in. And no one is any the wiser. Then suddenly, as the bidding war reaches the 50/51 stage, realization dawns. My friend is in the black! Think about it. Were the auction to end right there, he would, according to the rules of the game, already be a penny to the good. What a rip-off!

But the auction doesn't end there, of course. It runs and runs. In fact, it's not unusual for the £1 to go for £2, entailing a net profit on the initial investment of nearly £3 (£2 winning bid + £1.99 forfeit of the runner-up). What started out as good, old-fashioned greed soon metastasizes into a hideous, mutually *dependent* — yet at the same time mutually *exclusive* — foray into damage limitation. We don't just compete to maximize our gains. But also — who'd have thought it? — to maximize our *losses*.

DO UNTO OTHERS

In 1993, Manchester United won their first league title in twenty-six years. It was the first of eleven (and counting) under their legendary Scottish manager Sir Alex Ferguson. Ferguson is still in charge of Manchester United today and is now England's longest serving, and most successful, football coach. But in 1993 things were different. Up until that first league title, the Manchester United trophy cabinet had been gathering dust for a while — and Ferguson was worried that the arrival of a bit of silverware might go to the players' heads. What to do?

Some managers might well have let the players just get on with it, and bask in their hard-earned glory. Which, to be fair, Ferguson did. Up to a point. But the canny Glaswegian wasn't content with just the one title. He already had an eye on history. So he devised a plan to see them on their way: a simple stroke of genius that not only got the absolute best out of his players — it scared the shit out of them.

Ferguson remembers it like this:

> I said, "I've written three names down. I've put them in an envelope. Those are the three players that are going to let us down next season." And they're all looking at each other and saying, "Well, it's not me!" So the next season I did it again . . . *Of course, there was no envelope . . .* but it was just a challenge to them, because dealing with success is not easy.

Ferguson's strategy was lethal. Not only did Manchester United go on to win the league title again the following year, they have now, sixteen years on, chalked up a total of twenty-two major trophies under his stewardship. Success went to the players' heads all right — but in a manner that was constructive. They wanted *more*. And why? Because Ferguson released a virulent strain of persuasion that laid everyone equally low. That tapped straight into their ancient, hardwired need to be team members. And it was all just a simple con.

A policeman friend of mine uses a similar strategy in his job with problem kids. Richard Newman, who works as part of the Youth Offending Team in Cambridge, points out that teenagers are particularly susceptible to peer pressure and that the "you're ruining it for everyone else" line often succeeds where cajoling and coercion fail. To illustrate, he recalls an incident that took place several years ago on a group outing to the zoo.

> There were fifteen kids in the van and one of them, a really hard kid, wouldn't do up his seatbelt. "Gavin," I said, "do up your seatbelt — now!" He wouldn't listen. So I pulled over to the side of the road and told him I wanted to speak to him outside. "Come on then," he said, as he got out, "hit me!" I said, "Gavin, I'm not going to hit you. But I'll tell you something. We're not going anywhere until you do up your seatbelt."
> Then I pointed inside the van.
> "Now fourteen of your mates in there all want to go to the zoo," I said. "The longer we stand here talking, the less time we're going to have when we get there. So how about you just put on the belt and we get moving?" The result was almost instant. He thought about it for five seconds or so, then got back in. He was good as gold after that.

DYING TO BELONG

The wisdom of Newman's approach comes as no great surprise to his colleagues in law enforcement. True, not all teenagers are as easily led as others. And those that are not, a recent study shows, actually present with subtle neuroanatomical differences: decreased activity, on exposure to socially relevant stimuli, in the areas of the brain associated with motor preparation, planning, and attentional control (the right dorsal premotor cortex and the left dorsolateral prefrontal cortex), and greater functional connectivity between these regions and areas of the temporal cortex associated with action observation and processing. But the trend, one might argue, along with countless millions of parents, is pretty hard to ignore.

Young male syndrome is a pattern of behavior well known to forensic psychologists and police detectives alike. The population subset most likely either to kill or be killed comprises young men between their adolescence and midtwenties. Which, it turns out, is also the time of greatest competition for mates.

To casual observers the world over, it often seems inconceivable how murder and serious injury are inextricably linked to ostensibly trivial disputes. Yet we shouldn't be so surprised. There are evolutionary ley lines at work here, stretching all the way back from the pool tables and dance floors of inner city bars to the forests and savannahs of our past: a primordial whisper's ghostly intimation that it's not so much how we see ourselves that's important, but rather how others see us. And when one thinks about it, it makes perfect sense. Saturday night in the middle of London or New York would not, fundamentally, be all that dissimilar to that on the savannah of primal East Africa. More queues, I guess, but essentially the same dynamic.

In fact, in their midtwenties males are six times more likely than females to be the victims of homicide. Moreover — and herein lies the key — the majority of such homicides are committed in the presence of an audience. On the street. In a bar. In a club. They are, as it were, advertisements. But advertisements for what, exactly?

Some years ago now, I interviewed a sex offender who'd raped a woman at knifepoint. Two of his friends had also taken their turn. Why did you do it? I asked him. With the icy detachment typical of the psychopath, he shrugged.

"It's like buying drinks in a bar," he said. "You get a sense of identity. Of group camaraderie."

Such a view is echoed in the literature. While no one would deny — and we see it often enough — the insidious connection between group identity and violence, the polarity of such a dynamic can also be seen to reverse. Violent behavior may, in contrast, help facilitate group cohesion.

"One of the unique dynamics in gang rape," writes the clinical psychologist Nicholas Groth, "is the experience of rapport, fellowship and cooperation with the co-defenders. It appears [as if the offender] is using the victim as a vehicle for interacting with other men . . . behaving . . . in accordance with what he feels is expected of him . . . validating himself and participating in group activity."

A similar phenomenon may also be found in the gay community. "Suicide bumming" (as one guy I spoke to put it) refers to the practice by which healthy individuals deliberately have sex to contract the HIV virus. They achieve this by engaging in prearranged, systematic penetration with a succession of HIV-positive partners. One after the other. And then plug up the rectum to prevent the semen escaping. It's called, on the scene, "the gift."

In a bar in San Francisco, I ask one guy why he did it.

He has no hesitation in telling me.

"You feel, I don't know, more part of things," he says. "Like you belong more."

A friend — young, beautiful, and also HIV positive — comes over to join us. I ask him the same question. He smiles.

"It's a sign of commitment," he says. "Of solidarity. It's turning a negative into a positive. It's like getting a tattoo, only on the inside. It's like an immunological tattoo."

SUMMARY

In this chapter we've continued our tour of the key stimuli of influence by crossing the consciousness border. We entered the airspace of perception and social cognition and discovered that neural sophistication in no way precludes the kinds of instinctive, rapid response sets that we see in animals. Consciousness may well be handy, but it's slow: too slow, at times, for life to wait around for. So to bridge the gap, the brain employs heuristics — rules of thumb that rely on past experience; on learned associations between previously encountered stimuli. If consciousness had wheels there'd be eighteen of them. And a trailer so big you'd need planning permission to park it. Hardly ideal for getting across town in a hurry.

Taking the advice of a psychopathic genius con man (and why not?), we examined three areas of cognitive process: attention, approach, and affiliation. In each of these areas we saw how the brain may be brought to its knees as fast as any fixed action pattern we may find in the animal kingdom. We discovered, using some simple influence techniques, how we may harness the brain's impulsiveness to our benefit. And how, in the hands of the genius persuader, these same techniques can often cost us dearly.

In the next chapter, we widen the research spotlight. Having familiarized ourselves with the black arts of social influence, we now take in the white, and look at how psychological cat burglars such as lawyers, advertisers, salespeople, and cult leaders — those who work *with* us rather than against us — can crack our neural thought codes. It's easier than you think. The brain's security isn't exactly tight — and if you know what you're doing you can be in and out in seconds.

Memory Test

You have ten seconds to look at the words printed below. After the ten seconds have elapsed, turn over to p. 98 and answer the question at the bottom. Then turn back and read on for a debriefing . . .

SOUR   CANDY   SUGAR   BITTER   GOOD   TASTE
HONEY   SODA   TOOTH   NICE   PEA   CHOCOLATE
CAKE   HEART   TART   PIE

OK. How many of you said BOOMERANG? If you did, you're in the majority. But take another look and you'll find that the word SWEET isn't there either! The way it works is like this. The brain likes to impose order on the world. Make stuff nice and easy. When there are gaps it likes to fill them in; we get a sort of hunch, a kind of gut feeling about things.

Read over the list again and what do you notice? That's right, each word that appears on it is linked in some way to the word SWEET. Either semantically (e.g., CHOCOLATE) or linguistically (e.g., *SWEET*HEART) — which fools you into thinking that the word was there when it wasn't.

Your brain took a gamble, and ended up paying the price. Fortunately, on this occasion, it didn't prove too costly.

Which of the following four words did NOT appear on the list? SUGAR, TASTE, SWEET, BOOMERANG.

4

Persuasion Grandmasters

One morning, on arriving at his chambers, a lawyer discovers a surprise parcel waiting for him on his desk. On removing the outer packaging, he finds inside a box of the finest Havana cigars: a present from one of his clients for a particularly brilliant performance. On account of the rarity of the cigars, and of their not inconsiderable value, the lawyer decides to insure them. For $25,000. Over a period of the next few months, he then proceeds to smoke the cigars one by one (there are a dozen of them), until one evening, as he contentedly puffs on the last, he has an idea. Had he not insured the merchandise against precisely the fate that has now befallen them? Destruction, ahem, by fire?

Chancing his arm, the lawyer files a claim against the insurance company. Which the insurance company, perhaps not surprisingly, decides to contest. The case goes to court, and the lawyer — would you believe it? — emerges victorious. Even though the claim appears ludicrous, the judge remarks, there is nothing in the small print to preclude the award of damages. So a ruling is made in favor of the claimant. And the lawyer decamps some $25,000 better off. Nice work if you can get it.

Several weeks go by and the matter is quickly forgotten. Or so it seems. Then one morning everything changes. An envelope drops through the letterbox of the lawyer's chambers. From the insurance company. They are suing the lawyer for arson — on twelve counts — and a date has been set for the hearing. This time, of course, the boot's on the other foot. Pointing out that it would constitute an unprecedented breach of practice were the lawyer now to contradict the argument that had secured his victory in the previous case, the judge, on this occasion, awards damages, plus costs, to the insurance company. A sum total of $40,000. A case, one might say, of no smoke without fire.

Advertising may be described as the science of arresting human intelligence long enough to get money from it.

— STEPHEN BUTLER LEACOCK, *The Garden of Folly* (1924)

THE ART OF A GOOD STORY

What makes a good lawyer? I mean, a *really* good one. What's the difference between the guy who's brilliant in the courtroom and the guy who's just average? What have the stars got that the rest haven't? When I first started thinking about these questions I had no idea of the answer. But I knew a man who would.

Michael Mansfield is one of the world's greatest advocates. In his forty years at the bar, he's forged a reputation for fighting cases no other lawyer will touch. Consumed by abhorrence for hypocrisy and injustice, Mansfield's casebook reads like a résumé of modern British social history: the Bloody Sunday Inquiry, the Marchioness disaster, the Birmingham Six, Stephen Lawrence, Dodi Fayed and the Princess of Wales, and, most recently, the shooting of Jean Charles de Menezes.

I meet him at his chambers in Central London. He's a dapper sixty-seven with shoulder-length, sweptback hair and spacious, steely blue eyes. He's wearing a dark pinstripe suit and an open-necked shirt with a broad gingham check. The shirt is pink, and the hair is silver. Keith Barrett, I think. Only this time on the *right* side of the law.

I ask him what makes the great barrister great.

"Cases are won and lost not just on the strength of facts," he purrs, "but on impressions. A lot is achieved through the power of suggestion. The experienced barrister tells a story in court, and subtly spirits the jurors away with him on a narrative journey. The first thing members of a jury get in a courtroom is a gut instinct. They make up their minds with their hearts. The trick then is to present the evidence in such a way that it corroborates that initial gut instinct. It's just like in everyday life. It's much easier to convince someone that they were right all along than that they were wrong all along! Good barristers are also good psychologists. It's not just about presenting the evidence. It's about *how* you present it."

The importance of a coherent narrative is one of the fundamental principles of any type of persuasion. Not just in the courtroom but in the boardroom, on the campaign trail, or simply in everyday life. Frank Luntz is an American author and pollster, and a specialist in political persuasion. Early in his career, Luntz worked for the Independent candidate Ross Perot during his first presidential campaign — when Perot was at the peak of his powers, and the zenith of his popularity. Once, Luntz recalls, in Detroit, he organized a focus group to gauge the appeal of various Perot television ads. There were three of them in total: a biography, a Perot speech, and testimonials from other people.

When Luntz ran the ads in that order — biography, speech, testimonial — he found that Perot's popularity ratings within the focus group exactly mirrored polls taken from outside, among the general public at large.

No surprises there.

But when, by mistake, he ran the ads in the "wrong" order — testimonials, speech, biography — something very strange happened. Suddenly, the people in the focus group didn't really like Perot all that much. Divorced from his personal history, his opinions appeared intemperate. Which, counsels Luntz, just goes to show: "The order in which you give information determines how people think."

A friend of mine, Roz, provides a wonderful example of this "order effect" of persuasion. Roz's mother, Molly, is eighty-five and fiercely independent. For as long as I've known her, Molly has resisted the idea of acquiring even the most rudimentary assistance at home — despite the fact that she now has trouble dressing, and her memory is not what it was. Poor Roz. Countless times she'd pleaded with her mother to see reason. To at least consider the possibility of help, adding, "It'll make a real difference to your life — as it has to Mrs. McIntyre round the corner." But her efforts had come to nothing.

Then one day, just like Frank Luntz, she played the tape in the "wrong" order.

First, she mentioned that Kay [McIntyre] round the corner seemed much happier after the introduction of her homecare package. *Then* she suggested the possibility of her mum getting one, too.

The new storyline worked. Molly came round to the idea.

"It was like someone had waved a magic wand," Roz recalls. "Whereas before she'd been dead against it, out of the blue she just said, 'Hmmm. Well, I suppose we could give it a try. I guess I do need a bit of help in the mornings. And if it's working for Kay, it can't do any harm.'"

Kay was the undercoat on which the gloss of influence took hold.

Precisely how impressionable we are, just how swept along we can be by a good story, may be seen from the following. And remember, in the courtroom such judgments are crucial.

Have a quick read over the scenario below and then answer the question afterward:

> John is driving at 40 mph in a 30 mph zone and runs into another car at an intersection. The point of impact is on the driver's side. The driver of the other car receives multiple injuries including lacerations, a broken collarbone, and a fractured wrist. John himself is unscathed. John was speeding because he was rushing home to hide an anniversary present for his parents that he'd inadvertently left out on the kitchen table. The accident was exacerbated by the fact that he had to navigate an oil spill on his approach to the junction.

Question: On a scale of 1 to 10, where 1 = not to blame at all and 10 = totally to blame, to what extent do you think John was at fault in causing the accident? Indicate your answer by circling the scale below:

1   2   3   4   5   6   7   8   9   10
NOT TO BLAME AT ALL   TOTALLY TO BLAME

Now give the following scenario to a friend and ask them to make a similar judgment:

> John is driving at 40 mph in a 30 mph zone and runs into another car at an intersection. The point of impact is on the driver's side. The driver of the other car receives multiple injuries including lacerations, a broken collarbone, and a fractured wrist. John himself is unscathed. John was speeding because he was rushing home to hide a stash of cocaine from his parents that he'd inadvertently left out on the kitchen table. The accident was exacerbated by the fact that he had to navigate an oil spill on his approach to the junction.

Now I wouldn't mind betting that you and your friend had a bit of a disagreement over this one. And I'd also lay odds that your friend took a dimmer view of John's driving than you did. But hang on a minute. Why? *Why* is John more to blame when he's going home for the coke? Whatever John had inadvertently left out on the kitchen table, he was still doing 40 mph in a 30 mph zone, wasn't he?

At this point you may feel as if you've been set up. You haven't. Precisely these two scenarios were presented to a group of college students as part of an experiment conducted by the University of Ohio–based psychologist Mark Alicke. And you know what? Their responses exactly mirrored those of you and your friend. When John was on his way home to hide the anniversary present, the cause of the accident was split fifty-fifty between "something about John" and "something about the situation" (i.e., the oil spill). But when he was stashing the drugs it was a different story altogether. It was all *him*. Somehow, John's prior intentions render him a more "culpable" individual overall. And the more culpable we perceive someone to be, the more we attribute an internal, "dispositional" cause to their actions when those actions turn out badly.

JUDGING THE COVER BY THE BOOK

A man is in the final stages of getting things ready for his wedding. Everything is going well except for the one small matter of his bride-to-be's extremely hot younger sister. One afternoon, a week before the big day, he finds himself alone with her in the house. She sidles up beside him and suggests they go upstairs — before he finally settles down to a life of wedded bliss. The man begins to panic. Running through his options, he charges out of the house — only there, in the front garden, he discovers the rest of the girl's family all waiting for him. As soon as he emerges, they give him a big round of applause.

"Congratulations," says his future father-in-law. "You've passed the test. You have proved yourself a man of honor and integrity, and I'm delighted to give you my daughter's hand in marriage."

The fiancé can't believe it, and breathes a huge sigh of relief. His wife-to-be plants a big kiss on his cheek.

Moral of the story? Always leave your condoms in the car.

Within social psychology, the kind of error that you probably just made — and the kind of trap we fell into previously with John and his drugs and the anniversary present — supports nothing short of a cottage industry of research into the way our brains play tricks on us. The way, in little more than an instant, that the world's most complex computer can change, behind our very eyes, into the world's most complex whoopee cushion. Such cognitive flatulence — the irresistible tendency, when evaluating individual behavior, to give precedence to internal, dispositional factors over external, situational ones (especially when that behavior is our own and happens to be good, or is that of somebody else and happens to be bad) — has a name in psychology: the fundamental attribution error. And with good reason. It is, as its name suggests, fundamental.

Just how fundamental is revealed in a study conducted by Lee Ross, professor of social psychology at Stanford University. Pairs of college students first drew lots to determine who would play the role of question master and who would be the contestant in a mock quiz game. Each question master was then allotted a period of fifteen minutes in which to generate a series of "general world knowledge" questions. But there was a catch. These questions were to be specifically devised so that the answers were unlikely to be known by anyone else except the question masters themselves. True to form, once the quiz got underway, most of the contestants bombed. Big deal. Except that this wasn't the point of the study. Instead, at the conclusion of the quiz, everyone involved — the question masters, the contestants, plus some observers who hadn't actually participated in the game, just watched — were asked to estimate the level of each participant's general knowledge.

The results are shown in Figure 4.1. As can be seen, the question masters — either through feigned modesty or appropriate acknowledgment of the situational constraints — rated themselves as only slightly more knowledgeable than the contestants. Which may, or may not, have been true. But the contestants and observers — well, they were a different story altogether. Just look at the disparities in the middle and right-hand sections of the graph in relation to the knowledge base of the question masters and their "victims."

Even though the contestants had clearly overheard the researcher's instructions to the effect that the questions should be drawn from an idiosyncratic pool of knowledge unknown to anyone else but the question master . . . and even though they clearly recalled drawing lots to decide who would be question master and who the contestant so that, in a parallel universe, the roles might so easily have been reversed . . . even though they had experienced firsthand — and were perfectly well aware of — the overwhelming situational odds that were against them . . . they *still* exhibited a flagrant disregard for the impact that these odds might have had on the way things turned out.

Figure 4.1 — Actions speak louder than words — even when they're staged.

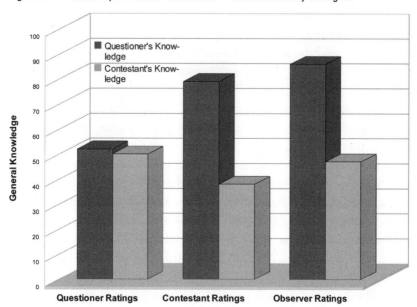

The question master *acted* smart. So the correspondent inference had to be that he or she *was* smart. In fact the observers rated the question master as being more clued up than 80 percent of all the other students at the university!

The fundamental attribution error offers us a prime example of what Michael Mansfield was referring to when he talked about impressions and the power of narrative. Take a rape case, for instance. In the courtroom, rape often constitutes a crucible of persuasion jujitsu in which opposing lawyers lock horns not so much over the minds of the jury as over their hearts. Let's take a look at each side in turn, beginning with the case for the prosecution.

The prosecutor knows that, for any behavior, an attribution of disposition (personal responsibility) overrides that of situation (external influences) when attention is focused on the agent of that behavior. This is what happened in the mock quiz game. Smart (the disposition of the question master, or agent) trumped experimental protocol (the situation in which the agent was acting), and did so quite spontaneously. We just can't help it. We have a powerful, inbuilt bias that predisposes us to think in a certain way: namely, that we do the things we do because we're the kinds of people who do those things! It's an evolutionary rule of thumb. A timesaving device programmed into our brains over millions and millions of years by natural selection. If, for every single behavior, we had no alternative but to engage in a protracted investigation of every possible contributing factor, how far do you think we'd get? Precisely. So we start with the person instead.

Experienced prosecutors are well aware of this principle. They are, exactly as Mansfield said, just as much psychologists as they are lawyers. So what do they do? What is their plan of attack? Well, it's like this. What the prosecutor will attempt to achieve is to focus the jury's attention solely on the alleged rapist. Maneuver them into a position where they're forced to ask themselves the question: *Why did he do it?* They'll concentrate on the defendant's previous relationships with women. (Perhaps he's exhibited aggressive tendencies in the past?) Or on his mental state at the time of the incident in question. (Perhaps he was drunk, or under the influence of drugs?) This, combined with an attendant emphasis on rape as a violent, as opposed to an erotic, act tells a simple, coherent "story" — one which plays right into the hands of the fundamental attribution error. With their attention focused solely on the defendant, and forced to account for his actions, there is, so far as the jury is concerned, only one reasonable conclusion. They'll presume that he is guilty.

In contrast, however, the case for the defense will endeavor to focus the jury's attention solely on the behavior of the victim. Get them to ask themselves the question: *Why was she raped?* They'll concentrate on factors such as how she was dressed. (Provocatively?) How she might have behaved prior to the attack. (Flirtatiously?) And on her previous sexual history. (Promiscuous?) This, combined with a counteremphasis on the possible erotic elements of rape, tells a totally different "story" — one likely to engender the antagonistic inference: she asked for it.

And in case you're wondering what previous sexual history has got to do with it — the answer, unfortunately, is quite a bit. Studies involving mock juries have shown that the amount of blame attributed to a rapist is often contingent just as much on victim characteristics as it is on the rape itself. A rapist is considered less culpable, for example, if he attacks a topless dancer than if he attacks a nun; and a woman who's divorced as opposed to one who's married.

Just like with John. And the drugs. And the anniversary present.

LOOK AT IT LIKE THIS

Following much media speculation over the demise of Osama bin Laden, bin Laden himself decides to send British Prime Minister David Cameron a letter in his own handwriting to show him he's still around. On opening the letter, Cameron discovers it contains the following cryptogram:

370HSSV0773H.

Despite hours of careful scrutiny, the British premier is completely stumped, so he types it out and mails it to his deputy, Nick Clegg.

Clegg and his aides are similarly none the wiser so they, in turn, send it to MI6. Again, no luck. So the message, over time, makes its way to MI5, to Scotland Yard, to the SAS, and eventually back to Downing Street. Still no one is able to decipher it. Finally, in desperation, a senior government official dispatches it to the CIA in Washington.

"This message," they say, "has remained inscrutable to the finest minds in Britain — as well as to David Cameron. As a last resort, we were hoping that you might be able to crack it."

Five minutes later a cable comes through to Downing Street: "Tell the Prime Minister to look at it upside down."

The psychological chicanery that we see in litigation is well known to students of social influence. It even has a name — *framing* — and is by no means confined to the courtroom. Sure, Mansfield and his buddies might well be the best in the business, but there are others out there who are just as quick on the draw. Advertising, politics, and sales, for instance, are just a few examples of professions in which this simple art of suggestion is also practiced. Then, as we just saw with the cryptogram, there's everyday life.

Psychologist George Bizer of Union College, New York, has studied the role of framing in politics. More specifically, in election campaigns. Do the ways in which voters come to express their views, Bizer wondered, impact on their strength of conviction?

To find out, Bizer asked a group of college students to read brief "news reports" of two fictitious candidates (Rick, a conservative, and Chris, a liberal). He then divided the students up into two groups. One group had to choose between the pair of statements "I support Rick" or "I oppose Rick." The other had to make an equivalent choice in relation to Chris. At this stage, each group also indicated their preferences for both candidates on a scale ranging from "strongly support" to "strongly oppose."

Then came the twist.

After registering their level of support for Rick and Chris, each group read a subsequent news report which was openly derogatory toward the policies of their preferred candidate. Allegiances were then reassessed.

Would the way in which participants initially framed their preferences for the two candidates have any bearing on how their opinions shifted during this second phase of voting?

The answer, it turned out, was yes.

In general, those students who'd conceptualized their preference for a candidate in terms of *opposition* to the other candidate (e.g., "I oppose Rick" rather than "I support Chris") were more resistant to change (i.e., were more likely to stand by Chris when he came under fire) than those whose preferences were *positively* framed.

"A simple change in framing, leading people to think of their evaluations in terms of whom they oppose instead of whom they support," says Bizer, "leads to stronger, more resistant opinions."

Framing, of course, doesn't just deal in emotions. It is, as Keith Barrett would say, primarily a virus of attention. To give you an example, ask a friend the following question:

How many liters of diesel does it take to fill up a jumbo jet? *Is it more or less than 500?*

Then ask another friend exactly the same question, only with a subtle twist:

How many liters of diesel does it take to fill up a jumbo jet? *Is it more or less than 500,000?*

Now ask each of them in turn to come up with a concrete estimate of just how many liters of diesel it really does take to fill up a jumbo jet. (Actually it's around 220,000.) Chances are you'll notice something rather interesting in their pattern of responses. The friend you asked second (more or less than 500,000) will come up with a higher estimate than the friend you asked first (more or less than 500).

The reason for this has to do with something called *anchoring*. What happens is this. Both friends quite literally use the numbers that you put into their heads (500 vs. 500,000) as frames of reference — anchoring points — on which to base their judgments. These numbers don't even need to be relevant to the issue at hand (in the jumbo jet example, we could just as easily have said 1 liter vs. 1,000,000 liters). They just need to be there. They persuade by presence alone.

In 2006, a team of German psychologists, Birte Englich, Fritz Strack, and Thomas Mussweiler, provided a classic demonstration of the power of anchoring — in the legal profession. The team took a group of experienced judges and asked them to read an outline of a case. The case involved a man who'd been convicted of rape. Once they'd familiarized themselves with the details, the judges were then divided up into two groups. One group was to imagine the following: that while the court was adjourned, they received, in their chambers, a telephone call from a journalist. This journalist posed them the following question: Would the sentence be higher or lower than three years? The other group was presented with a slightly different scenario. They, too, were told that they'd receive a call from a journalist — only in this case the journalist would inquire whether the sentence would be higher or lower than *one* year.

Would this simple differential in numbers — three versus one — have any effect on the *actual* number of years meted out by the judges?

It sure did. Just as the anchoring hypothesis predicted, the average length of sentence handed down by the judges in the first group was thirty-three months. In the second, it was twenty-five.

DEVIL IN THE DETAIL

The most obvious place we might encounter anchoring is, of course, in the marketplace. We've all haggled over the price of something or other — and opened with a figure we know we'll have to improve on. But what isn't so obvious is the ease with which some of the more sophisticated persuasion strategies — techniques we barely even notice — can pickpocket our brains. Take pricing, for instance. Ever wondered why your favorite hair product is marked up at £9.95 instead of a round £10? Chris Janiszewski and Dan Uy at the University of Florida have recently contemplated exactly such a question — and arrived at a surprising conclusion. Rather than "nine just seeming cheaper than ten" (which is what most people say when asked), it's actually a bit more complicated.

Janiszewski and Uy performed a series of experiments in which volunteers were presented with hypothetical salesroom scenarios. In each scenario, volunteers imagined that they were buying an item (e.g., a high-definition plasma TV) at a particular *retail* price. They were then required to guesstimate the *wholesale* cost to the vendor. "Consumers" were divided up into three groups. One group of buyers was told that the TV cost $5,000; a second group that it cost $4,988; and a third that it cost $5,012.

Would these different anchoring points — negligible in relation to the total price — have an impact on participants' wholesale quotes?

Remarkably, they did.

Those buyers who were given the $5,000 price tag estimated wholesale costs to be significantly lower than those who'd deliberated over the more precise figures. Moreover, recipients of the $5,000 tag showed a far greater propensity to estimate a wholesale cost in round numbers than those who'd started without them.

But why?

To explain their results, Janiszewski and Uy speculate as to what the brain might be doing when it calculates such differentials — the precise anatomy of the comparison procedure. Or, more specifically, its units of measurement. Could it be that these units of measurement are variable, and contingent on certain characteristics of the initial price? Let's say, for example, that we go into a shop and spot a clock radio on display for £30. On seeing the radio we might well think to ourselves: that radio is really worth around £28 or £29. *Whole numbers.* On the other hand, if we see it retailing at £29.95 we might *still* believe that it's worth less than the asking price — but the yardstick we use to evaluate the disparity is different. This time its intervals are smaller. Rather than thinking in whole, round pounds we think, instead, in loose change. We consider, perhaps, £29.75 or £29.50 as the "true" wholesale value — less of a differential than if we were thinking in whole numbers. Which makes it more of a bargain.

To test their theory, Janiszewski and Uy headed out of the lab and into the real world. To Alachua County, Florida, to be precise. There, they looked at real estate — comparing the asking prices of houses with the amounts they actually sold for. Just as they'd predicted, sellers who put their houses on the market for more precise sums (e.g., $596,500 as opposed to $600,000) got consistently closer to the asking price than round-number vendors. And that wasn't all. In the event of a market slump, homes advertised for round-figure sums showed a higher rate of depreciation than those with more "precise" price tags. Over as brief a time span as just a couple of months.

ENWRAPTURE

In more visceral modes of advertising, such as that for food and drink, framing and logical inference often part company completely. Here, the corporate machinations of spin hit directly on neurophysiology—precipitating sneaky, unconscious shifts in low-level sensory perception. Take the drinks industry, for instance. Cheskin, a market research organization based in Redwood Shores, California, experimented with different colored backgrounds on cans of 7Up. Some cans were more yellow. Others more green. In both cases, the beverage inside was the same. If customers hadn't gone nuts, bosses back at 7Up HQ might well have seen the funny side. Those who purchased the yellow cans reported an unfamiliar "lemony" flavor to the contents, while those who bought the green cans complained—you got it—that there was too much lime in the mix.

"When we decide in the blink of an eye whether or not food tastes good," says Cheskin CEO Darryl K. Rhea, "we're reacting not only to the evidence from our taste buds and salivary glands, but also to the evidence of our eyes, memories, and imaginations."

David Deal, creative director at Deal Design Group in San Diego, agrees. Imagine you're throwing a party. We've all done it—grabbed a bottle of vodka we've never even seen before retailing at £30, when sitting there right next to it is a more familiar brand for £10. But why? Do we really think we're going to taste the difference? I don't think so. To me, vodka tastes of nothing. Can nothing—even at £30 a bottle—be made to taste of *something*?

Deal's got the answer—and it has nothing to do with taste. Instead, it's about *feel*. About what's known, in the trade, as *emotional branding*.

"They're selling you the experience of being at a chic party and drinking a martini made with vodka that was carved out of ice in the depths of Finland," he explains. "If you put a beautiful bottle next to a consumer, they'll say it contributes to the taste."

Which explains, according to Washington-based industry group Point of Purchase Advertising International, why up to 72 percent of our purchase choices are made right there and then. On the spot. Spontaneously. And why a simple change of color, or a wrong choice of word, can just as easily turn us *off* a brand as *on* to it.

British entrepreneur Gerald Ratner, founder of Ratner's jewelry chain in the U.K., famously watched his multimillion-pound business go down the drain after a disastrous quip at a meeting of the Institute of Directors. Ratner — the aptly named Sultan of Bling — offered the following, canny insight into why the stuff in his shops was so cheap.

"Because it's crap."

He further elucidated — astutely, eloquently, and with terminal disregard for the art of obfuscation — on the quality of some of his earrings: "Cheaper than an M & S prawn sandwich but probably wouldn't last as long."

Ratner's shares plummeted. Not because some dark, elusive secret had finally slipped out. (Anyone who didn't already know that 18 carat gold, diamond-encrusted watches usually went for more than £19.99 needed their heads tested.)

No. It wasn't because everybody suddenly knew that what they were buying was crap.

It was because everybody suddenly knew *that everybody else knew* that what they were buying was crap.

GIVE AND TAKE

Framing and anchoring are just two of the techniques that can increase persuasive power. There are others — as "cold caller" Pat Reynolds knows only too well.

In his first week of joining a telesales company, Pat Reynolds chucked everything he'd been taught into a big file called "bollocks" (his words, not mine) and instead developed his own, somewhat unique, style of sales pitch. Over the last few years this pitch has earned him a BMW Z4 Roadster, a pilot's license for a light aircraft (not cheap), and a substantial down payment on an apartment. The company he works for deals in building and renovation work. His secret? A demonic combination of making people laugh — while, at the same time, actively courting *rejection*. Here, in Pat's own words, is how it works:

> People call it cold calling but if I do my job properly only one in ten calls is *really* cold. I begin by making them laugh. "Are you superstitious?" is one of the things I ask. If someone rang you up out of the blue and asked you that, you'd be curious, right? At least, you'd be less likely

to put the phone down than if I said, "This is Joe Bloggs calling from such and such."

That's the first key. You've got to keep them on the line. You can't sell anything to a dial tone. So most people say no, they're not superstitious. So then I say, "Well, will you give me £13.13?" Nine times out of ten this gets a reaction. Usually they laugh and say, "Who *is* this?"

And then I'm in. But I don't try to sell them anything. That's a no-brainer. I do the total opposite. I say, "Look, I know you want to watch *EastEnders*" (or *Coronation Street,* it's good to call around a quarter of an hour before the soaps start so you can use that line: it makes them think you're just like them, always good for a sale) "and I know you probably don't really want any building work done, but do you know anyone — any friends, or family, or friends of friends — who might?"

Because I've made them laugh, and because they also think I've done them a favor by letting them off the hard sell, they usually give me a couple of names or ask me to ring back after they've made some calls. And I make a point of asking them if that's OK. I say, "Is it OK if I call you back?" And they say, "Yes." Sounds trivial but that's important. It makes it like a contract. Kind of cements the deal — like a verbal handshake.

Great, isn't it? After only two or three minutes on the phone, *they're* working for *me*! Maybe I should think about giving them a cut. So the next call you make isn't really cold anymore. It's a referral. One good turn deserves another and all that.

Strictly speaking, Pat Reynolds belongs in the previous chapter. Or does he? I can't make up my mind whether he's a first-rate con man or just extraordinarily good at what he does. Maybe it's a bit of both. Whichever it is, his strategy for drumming up business brings to the table a completely new style of framing — a virus not of attention, but of approach. A virus as endemic to the sales industry as it is, unfortunately, to my better half: *emotional blackmail.*

In sales (as opposed to marriage), emotional blackmail needs to be subtle. Preaching, lecturing, pleading, and bullying are about as useful in the showroom as a snooze button on a smoke alarm. Instead, just like Pat Reynolds, the successful salesperson treads carefully.

Just look, for a moment, at the way Reynolds operates.

For all his bravado and roguish streetwise charm, he's a serious player.

Not everyone who works in a call center learns to fly on the proceeds. Nor do they roll off a gleaming garage forecourt in a top-of-the-range convertible. He makes an awful lot of money where an awful lot of people simply go under. And how? By reverting to first principles. To a time when persuasion had yet to hit on language. By releasing from the depths of human evolution one of the most powerful genies of influence known to man: the principle of *reciprocity*.

Robert Cialdini, Regents' professor of psychology and marketing at Arizona State University, has shown precisely how powerful the pull of reciprocity is, how rightful its place in the arsenal of the elite persuader, in a study that (on the surface, at least) looked at individual differences in altruism. In fact, it was really about compliance.

Here's how it worked. First, Cialdini and his colleagues stopped random passersby in the street and divided them up into two groups. Each group was then posed a question. Those in the first group were asked how willing they would be to supervise a cohort of inmates from a juvenile detention center on a day-trip to the zoo. Oddly enough, few of them were interested. Just 17 percent. For those in the second group, however, the furniture of influence was subtly rearranged. These participants were first asked a rather different question: would they consider putting in two hours a week as a volunteer counselor at the detention center for the next couple of years? This time, surprise, surprise, there were no takers at all.

But then something rather spooky happened.

When Cialdini and his coinvestigators responded to such refusals with the rejoinder, "OK, if you're not willing to be a counselor would you be prepared to supervise a cohort of inmates from a juvenile detention center on a day-trip to the zoo?" — precisely the same question as that put to the first group — agreement to the request shot up to 50 percent. It pretty much tripled the previous consent rate.

Of course, it doesn't take a genius to work out what's going on here. The power of reciprocity, concluded Cialdini, extends far beyond the distribution of gifts and favors. It may also be applied to the kinds of *concessions* we make to one another. If you reject my larger request and I, ostensibly, make a concession by retreating from that larger request to a smaller one, then you're likely to concede in kind by "meeting me halfway."

Which means, if what I've been after all along is merely your agreement to the smaller request, that I've got what I wanted. Right?

CONTRACTS OF MIND

Pat Reynolds's deployment of the reciprocity principle is, from a scientific perspective, little short of perfection. He couldn't execute better were he, like Cialdini & Co., part of a carefully controlled psychology experiment. Because he spares his "customers" the hard sell — empathizing with their predicament of being called up just as their favorite TV soap is about to start, and asking "considerately," apologetically even, if they might know of anyone else who'd be interested in his services — they feel obligated to give him a name. But what they conveniently overlook (from Reynolds's point of view) is that, far from being chivalrous, he's actually a pain in the arse — the *chef de cuisine* of a two-tiered trifle of torment: calling them up in the first place, and then, during the commercial break or whenever, prising them out of their armchairs to ransack address books and rummage around for contacts.

And the story doesn't end there. Reciprocity, it turns out, isn't the only virus of approach that Reynolds incubates in his clientele. There's another: *cognitive consistency.* The impromptu request for permission to call back later is, as Reynolds himself points out, nowhere near as casual as it seems. Far from it. It's an ancient maneuver of pure, primeval persuasion. A trigger, should permission to call back be granted, for the clients to stick to their word — to honor *their* side of the "bargain" and dutifully drum up business.

And it works. Quickly. Covertly. Subcortically. In fact, to compliance connoisseurs, reciprocity plus consistency is a pretty standard concoction. The two ingredients often go hand in hand. If the evolutionary origins of reciprocity may be traced to the division of labor and the facilitation of ingroup cohesion (hunting, transporting large objects, and constructing shelters all involve teamwork), then the psychological properties of consistency and commitment may be seen as a kind of "entry card" to such inclusion. These are the attributes that ensure dependability. That signal to the group that we *are* as good as our word.

Just how much influence the desire to appear consistent can wield over our behavior, and just how persuasive someone can become if they contrive to hack into such ancient evolutionary frequencies, may be seen from the following: a dazzling example of subterranean persuasion from Chicago restaurateur Gordon Sinclair. In the late 1990s, Sinclair was having trouble with no-shows. It's something that every restaurateur has to deal with, one of the real downsides of the business—a customer makes a booking over the telephone and then, without prior warning, fails to turn up. As things stood at the time, the no-call/no-show rate in Sinclair's restaurant hovered around 30 percent. But, at a stroke, he slashed it to 10.

The key to the problem, Sinclair discovered, lay in what his receptionist said on the telephone. Or, more accurately perhaps, what she didn't say. Prior to intervention, the receptionist, whenever a customer called to make a booking, would issue the following INSTRUCTION: *Please call if you have to change your plans.*

Subsequent to intervention, however, she amended this slightly to the following simple REQUEST: *WILL YOU please call if you have to change your plans?*

And then she would pause and wait for the caller to answer.

Just those two extra words, plus that all-important pause, changed the whole nature of the problem.

Why?

Because the question begged an answer, and the silence—as all silences do over the telephone—needed filling. By answering "Yes" to the question, *"Will you please call if you have to change your plans?"* callers were handing themselves a memorable psychological grid-reference: a contractual landmark on which to fix a bearing. Future action suddenly became illuminated by previous commitment. And, once committed, the locus of responsibility subtly shifted. Rather than face the prospect of letting just the restaurant down, customers now faced the prospect of letting themselves down, too.

Sinclair's technique has a name in the influence literature—foot-in-the-door—and was officially unveiled back in 1966 in an experiment so outlandish that the researchers, Jonathan Freedman and Scott Fraser, achieved a rare feat in the annals of scientific enquiry: they dumbfounded even themselves. The ball started rolling in a smart, affluent neighborhood of Palo Alto, California. An associate of the researchers, posing as a volunteer worker, began knocking on doors and assailing residents with a quite extraordinary proposal: the erection, slap bang in the middle of their front lawns, of a gargantuan public-service billboard depicting the words DRIVE CAREFULLY.

To facilitate the decision-making process, residents were shown a picture of the setup—where it would go, how it would look—and it wasn't pretty. The billboard was as big as the house, and took up most of the lawn. Not surprisingly, most of them (73 percent) told the researchers, in no uncertain terms, exactly what they could do with it. And that wasn't pretty either. Except, that is, for one particular group: 76 percent of which actually *acquiesced* to the installation of the billboard. Can you believe that? Neither could I when I read it. So what was so special about this one particular group? Were they mad? Were they bribed? What could have induced them to take such leave of their senses? To pack up their sprinklers and kiss their hydrangeas goodbye?

The answer is actually very simple. Two weeks prior to the visit about the billboard, a different "volunteer" had also appeared at their doors. On this occasion, however, the request had been relatively harmless: the placing, in residents' front windows, of a three-inch-square sign bearing the words, BE A SAFE DRIVER. No problem there whatsoever. In fact, so insignificant had been this request, and so in tune with neighborly sentiment, that almost *everyone* approached had consented.

Yet how costly it had proved in the long run. That one insignificant request — once agreed to and long since forgotten — had initiated the ripples of a murderous commitment tsunami — had precipitated acquiescence to a consonant, yet far greater, demand that awaited them just round the corner: the subsequent display of a billboard a hundred times bigger.

Double-glazing salesmen the world over suddenly sat up and took notice. The cat was now out of the bag — sporting snakeskin boots and a beat-up, ten-gallon hat. To get a person to sign away his life, you do it a day at a time.

HARD TO GET

In sales, a related technique to that of foot-in-the-door is known as *lowballing*. I first encountered low-balling in a job I took as a student selling televisions. The routine would go as follows. Customer walks in through door and shuffles up to Set. Assistant — quite possibly me — approaches Customer and, after preliminary small talk about Weather, offers him a Deal. Deal considerably undercuts those offered by other showrooms in the neighborhood and Customer jumps at it. But Manager has other ideas. Unbeknownst to Customer he has no intention of honoring it. None of us do. Not at the price initially quoted, at any rate. It's a lure. A piece of psychological stage magic to entice Customer to *decide* to make the purchase.

But this is just for starters. There then ensues the Byzantine Closing Ritual — procedures designed for no other reason than to make the decision stick: the completion of Complicated Contract; the talking through (as exhaustively as possible) of Labyrinthine Financial Agreement; and the generous encouragement of Customer — "Try it, you'll like it" — to take the television home for a trial period: "Just to see if it fits in with the decor, you know, that sort of thing." Which, of course, it does. Eleven times out of ten.

Can you see what's happening here? The greater the number of hoops the customer is made to jump through, the greater becomes their commitment to the cause. A particular favorite of the guy I used to work for was to find out how many forms of ID the customer had on him in the shop, and then to insist, irrespective of how many it turned out to be, that he still required one extra: "With the increase in security, it's the law."* Undeterred by this additional encumbrance — indeed, spurred on by the generous terms agreed at the outset — the customer would then be dispatched to procure the requisite documentation, none the wiser that at the very moment he walked out of the shop, free will hurled itself under the wheels of an impending direct-debit agreement.

Inevitably, of course, somewhere along the line, something would then "come up." The original price did not include tax; I, or some other bungling member of the sales team had "made a mistake" which the manager — "Sorry, my hands are tied" — just couldn't allow to stand. But do you think it made any difference? Put a dent in our chances of closing? Did it hell. In the overwhelming majority of cases the customer would *still* walk away with the television even when the price ceased to be competitive and actually rose *above* that offered by other showrooms. Every dotted line, every extra form of ID, every smile, every handshake, made it more and more worth having.

In the end, the thought of getting their hands on that television was the only thing left in many customers' minds. They couldn't, for want of a better phrase, switch it off.

LINGUISTIC ASSASSINATION

Mary, a pious do-gooder and local church gossip, was always sticking her nose into other people's business. She wasn't overly popular with the other church members, but she had a formidable reputation and no one wanted to mess with her. One day, however, after noticing his pickup van parked outside a bar, she accused Bill, a newcomer, of having a drink problem. Indeed, at the church's next committee meeting she made it quite clear, to Bill

---

*Sometimes it got silly. One guy had *six*.

and to everyone else, that it was an open-and-shut case: anyone who saw his van parked there would know exactly what he was up to. What other explanation could there be?

Bill, a man of few words, looked at her for a moment, and then left. He made no attempt to explain himself, nor did he deny that what Mary had seen was true. He said nothing.

Later that evening, Bill quietly parked his van in front of Mary's house. And left it there all night.

Language evolved as an aid to communication. But now that we've got it have you ever noticed how it's often the things we *don't* say that turn out to be most important? How, in the right — or, indeed, wrong — hands just the one word (sometimes less!), discreetly and expertly placed, can make all the difference?

The American political drama *The West Wing* once featured an episode in which the Democrats wanted to call Arnold Vinick (the Republican candidate for president) old. The problem, however, was that they didn't want to do so directly. An overt, premeditated attack on Vinick's advancing years would not only have been counterproductive to Democrat nominee Matt Santos's campaign, there was also the danger that it might not have "stuck." What to do?

The solution comes in a brief but hilarious dialogue between Lou (Santos's director of communications) and Josh (his deputy chief of staff). Having rejected the irony of labeling Vinick "vigorous," Lou, without thinking, refers to him as "spry." Inconveniently spry, for Santos's political ambitions. Josh seizes on the opportunity. Whoever heard of anyone under the age of seventy being called "spry"? he muses. Though, on the face of it, flattering, spry is just one of those words: what the large print giveth the small print taketh away. It says "old guy versus young guy" without so much as the slightest nod to age — the perfect disguise for a pop at the Republican feeb!

Words like "spry" are great, aren't they? They say one thing, but allude to something completely different. Josh is right. While technically a compliment, spry is a label that is really only applied to the elderly. And so, as such, is exactly what Matt Santos's campaign team is looking for. By referring to Arnold Vinick as "spry" (in much the same way that Barack Obama referred to John McCain's "half century of service" in the run-up to the 2009 U.S. election), they can draw the public's attention to their opponent's age, while at the same time appearing evenhanded: a backhanded insult and political gift horse in one.

In 1946, Solomon Asch — whose work on conformity we encountered in the previous chapter — demonstrated precisely how language can color our social perception in an experiment now regarded as *the* classic study of the way we form impressions. First, Asch presented participants with a list of trait descriptions. These descriptions all related — supposedly, at least — to the same individual. But there was a catch. Prior to drawing up the list he divided the participants into two groups — and then adjusted the list accordingly so that each group received exactly the same descriptions but for one (as it turned out, crucial) variation.

One group was presented with the following list . . .

intelligent, skillful, industrious, warm, determined, practical, cautious

. . . while the second group received this one . . .

intelligent, skillful, industrious, cold, determined, practical, cautious.

Spot the difference? The lists are indistinguishable except for the words "warm" and "cold" fiendishly sandwiched in the middle of either one.

Having been allocated their respective lists, Asch then asked both groups to use the descriptions to guide them in their choice of further traits — selected from a supplementary list — which they considered "went" with the personality profile initially presented.

Would the inclusion of the simple "warm-cold" differential, Asch wanted to know, be sufficient to cause a disparity in each group's choice of attributes?

The answer, without a shadow of doubt, was yes.

The group that received the "warm" list co-opted, from the supplementary inventory of traits, descriptions such as "happy" and "generous."

In contrast, however, the group that received the "cold" list selected traits such as "calculating" and "unsympathetic:" unflattering, to say the least, compared to their virtually identical counterparts.*

---

*This is a study that you can easily perform on your friends. The complete list of supplementary trait adjectives used by Asch in the original version of his task is provided in Appendix 2. Try the original version first and then experiment by varying the content of both the primary (e.g., warm/cold) and secondary (e.g., intelligent, practical) trait lists to see which descriptions really do make a difference.

LIVING IN OUR OWN LITTLE WORDS

A man is walking through the zoo one day when he sees a little girl leaning into the African lion's cage. Suddenly, the lion grabs the little girl by her jacket and tries to pull her inside. Her parents start screaming hysterically.

The man runs to the cage and whacks the lion on the nose with his umbrella. Whimpering from the pain, the lion retreats and lets go of the little girl. The man reunites her with her parents — who thank him repeatedly for saving their daughter's life.

Unbeknownst to the man, a journalist has been watching what happened.

"Sir," he says, walking up to him afterward, "that was the bravest thing I ever saw in my life."

The man shrugs.

"It was nothing," he says. "The lion was in a cage and I knew God would protect me, just as He protected Daniel in the lion's den. When I saw the little girl was in danger, I just did what I thought was right."

The reporter is gobsmacked.

"Is that a Bible I see in your pocket?" he asks.

"Yes," says the man. "I'm a Christian. In fact, I'm on my way to Bible class right now."

"I'm a journalist," replies the reporter. "And you know what? I'm going to run what you did on tomorrow's front page. I'm going to make absolutely certain that your selfless act of heroism doesn't go unnoticed."

The following morning the man buys the paper.

The headline reads as follows: RIGHTWING CHRISTIAN FUNDAMENTALIST ASSAULTS AFRICAN IMMIGRANT AND STEALS HIS LUNCH.

Words, it should be becoming clear by now, are psychoactive. Taken aurally or visually and delivered to the brain in milliseconds, they can modify our thought patterns and influence the way we approach situations just as fast as anything we may be able to pick up on a street corner.

In the media, for example, the right kind of word can grab our attention, can fire our emotions, just as easily as the biggest billboard. A case in point is the rise of political correctness. In 2005, the Global Language Monitor — a nonprofit organization that does exactly what its name suggests — issued a tongue-in-cheek list of the year's most politically correct words and phrases. Top of this list was the term "misguided criminals" — an elegant euphemism rustled up by a BBC commentator in the wake of the London bus and tube bombings. Over fifty innocent civilians had perished in the blast. But "terrorists," apparently, was deemed too emotive.

Also on the list were the terms "thought shower" which, in deference to those with epilepsy, had apparently, on one occasion, replaced "brain-storm." And "deferred success" which on another had substituted for "fail-ure." Then, of course, there was "womyn."

It's difficult in the face of such buffoonery to countenance the possibility that there may be a serious side to all of this. Yet there is. In the law courts, for instance, the hypnotic potential of language is recognized only too well as an impish impediment to justice. This is precisely the reason why, in cross-examination, "leading" questions are so vehemently overruled.

A classic study conducted in 1974 by Elizabeth Loftus of the University of Washington and her colleague John Palmer provides clear evidence as to why this should continue to be the case. The centerpiece of the study comprised a video clip of a minor road traffic accident — a moving car making contact with a stationary one — which Loftus and Palmer played to two groups of participants. After they'd watched the video, the researchers then posed the same question to each group. At what speed did the moving car run into the parked one? Surprisingly, in spite of the fact that both groups had seen exactly the same clip, they gave radically different answers. The response of one group (averaged across its members) was 31.8 mph, whereas that of the other group (similarly averaged) was 40.5 mph.* How could this be?

The reason, of course, was simple. On posing the question, Loftus and Palmer used subtly different wording.

Of one group they asked: "At what speed did Car 1 *contact* Car 2?"

While of the other group they asked: "At what speed did Car 1 *smash into* Car 2?"

That insidious one-word disparity between the two questions made all the difference to the answers. Not only that, but those witnesses who'd been asked about the *smash* reported glimpsing broken glass at the scene of the accident — even though, in the original video clip, there wasn't any. No won-der that "leading the witness" elicits such urgent and vociferous objections.

---

*The actual speed with which the moving car made contact with the stationary one was 12 mph.

But it's not just in the courtroom that we need to be on our guard. Similar effects to those documented by Loftus and Palmer are also found in politics. As I write, Barack Obama's honeymoon period as President on Fire is well and truly over; he didn't go to Berlin for the twentieth anniversary of the wall coming down; has become too preoccupied with health care reform — and too blasé about unemployment; bowed too low before the Japanese emperor; and permitted the Chinese president to bar questions at their joint press conference. Democrats, of course, traditionally have a hard first year in office. In politics, the worst kind of devil is the one that's in the detail — and of the past seven presidents, Bushes I and II rank top in popularity after their inaugural year in power, and Clinton and Obama bottom. Moreover, you might ask, which U.S. president ever got *anything* out of the Chinese? Yet Obama, even before he came to the White House, wasn't averse to throwing political curveballs. During the election campaign of 2008, records show that despite enjoying near unanimous support from African Americans in the polls, he steadfastly resisted all efforts to label himself as a "black candidate."

"I reject a politics that is based solely on racial identity, gender identity, sexual orientation or victimhood generally," Obama asserts in *The Audacity of Hope* — a line he took right from the very beginning of his campaign.

But why? Why, when you have the overwhelming backing of a significant portion of the population, not take advantage of your ethnicity? Appeal to the sense of history that hunkers down in all of us?

The answer, as *Time* columnist David von Drehle rightly points out, is that "black" is just one of those words. One of those "central traits" like "warm" and "cold" that Solomon Asch first started playing around with back in the 1940s.

"As soon as the race label is added," writes von Drehle, "some of the audience tunes out, others are turned off and still others leap to conclusions about who you are and how you think. Obama has written that race was his 'obsession' growing up but that he long ago left that burden behind. Now he lays claim to the whole spectrum: 'the son of a black man from Kenya and a white woman from Kansas' with 'brothers, sisters, nieces, nephews, uncles and cousins, of every race and every hue, scattered across three continents.'"

True, maybe. But expedient as well.

BOTTOM LINE

Frank Luntz, who we met earlier in this chapter, runs focus groups specifically designed to unlock the power of language. To uncover the perfect phrase — a golden word, or grouping of words, whose meaning remains untarnished by interpretation. The zeitgeist zinger that says exactly what voters hear — as well as what the politicians want them to hear. A semantic chord that resonates precision.

Luntz is a linguistic scuba diver — a pirate of the idiomatic unconscious. First, he throws out buzzwords — key political terms or familiar, topical sound bites — to which members of his focus groups free-associate. These free-associations factorize into second-order words or phrases that then form the basis of subsequent group discussion. From these discussions, Luntz filtrates a third generation of words. Then, after further rounds of debate, a fourth or a fifth. The eventual result is a distillate of meaning several times removed from the word or phrase originally presented, yet deeply imbued with primal connotation. A new word or phrase that means, quite literally, what it says.

Back in 2000, *New Yorker* correspondent Nicholas Lemann attended one of Frank Luntz's focus groups and witnessed firsthand his trademark linguistic alchemy. The show kicked off with the word "government." What, Luntz asked those present, did that word mean to them? Responses — at first — were nothing to write home about. "Controlling," "laws," "security," "bureaucracy," "corruption" . . . no surprises there. Then one of the assembled, a general contractor, blurted out the following: "A lot of regulations . . . a lot of stuff that I don't need to put up with. They could leave me alone a little bit. I would be a bigger company if I could have two things: a little less law, and a little more help."

Bam! Things had just kicked up a notch. Here was something that Luntz could really run with — and with a judo master's skill for spotting that crucial opening, he seized on it.

He turned to the rest of the group. "What's your reaction to that?" he asked them. A little while later, after a prolonged invective against laws, politicians, and — hey — the whole Washington thing in general, he was chalking up five key words on an easel. Opportunity. Community. Responsibility. Accountability. Society.

Luntz probed deeper. When the group reflected on their core values, the things that really mattered to them in their lives, which of the five was the most important? A show of hands quickly revealed the answer. Opportunity came out on top. Accountability second. And community last.

But what, exactly, did that word "opportunity" mean to people? Luntz asked.

As the group yelled out answers — "right to choose," "personal control," "no obstacles," "everyone gets a chance," "founding principle of the country" — he hauled a fresh sheet of paper onto the easel.

Once again, a pecking order was established as the relative importance of these core atoms of democracy, these mitochondria of liberty, was put to the vote. This time, "founding principle" won gold, "everyone gets a chance" got silver, and "right to choose" the bronze.

Luntz then turned to Lemann.

"You have the Republican and Democratic definitions of opportunity right there," he pronounced. "The Republican is 'right to choose' and the Democratic is 'everyone gets a chance.' Individual versus global."

Luntz's philosophy is, of course, hardly new. In the early 1900s it was Theodore Roosevelt who first articulated the benefits of mind reading in the political arena. "The most successful politician," Roosevelt observed, "is he who says what the people are thinking most often in the loudest voice." Or, as someone else more succinctly put it: the best way to ride a horse is in the direction in which it is going. But what Luntz does do is drag Roosevelt into the modern era, using political insight and discursive psychology as a backstage pass to the brain, as a linguistic MRI scan of affective, semantic form — revealing the hairline fractures and microscopic lesions to which all communication is subject.

Take the phrase "oil drilling," for instance. If you think it couldn't be clearer, think again. In 2007, Luntz got a group of people together and showed them a picture of a deep-sea drilling project in the Gulf of Mexico. He then asked the group how they viewed the picture. Did it "look like exploration or drilling?" Incredibly, 90 percent of those questioned said it looked like exploration.

"If the public says after looking at the pictures, 'that doesn't look like my definition of drilling, it looks like my definition of exploring,'" Luntz argued, "then don't you think we should be calling it what people see it to be, rather than adding a political aspect to it all? . . . Drilling suggests that oil is pouring into the ocean. In Katrina, not a single drop of oil spilled in the

Gulf of Mexico from the rigs themselves. That's why deep-sea exploration is a more appropriate term."

Back in 2002 — at a time, by his own admission, when the scientific evidence was perhaps not as compelling as it is now — Luntz prescribed a similar kind of rehab for the term "global warming." In a memo to George W. Bush entitled "The Environment: A Cleaner, Safer, Healthier America," he wrote as follows:

> The scientific debate is closing [against us] but not yet closed. There is still a window of opportunity to challenge the science. . . . Voters believe that there is no consensus about global warming within the scientific community. Should the public come to believe that the scientific issues are settled, their views about global warming will change accordingly. Therefore, you need to continue to make the lack of scientific certainty a primary issue in the debate, and defer to scientists and other experts in the field.

Result? Global warming slipped into something more comfortable. Something less alarmist, less loaded — and, politically at least, more savvy. Something — you may have heard of it — called "climate change."*

SUMMARY

This chapter, in theme if not in content, follows on pretty much from where we left off in Chapter 3. There, if you recall, psychopathic genius and natural-born persuader Keith Barrett introduced us to what he referred to as the Three A's of social influence — attention, approach, and affiliation — and we examined how his typology, gleaned from a lifetime of thinking on his feet, stood up to the rigors of scientific scrutiny. Not too badly, as it turned out.

Here, in Chapter 4, we opened the door a bit wider. We stepped out of the shadows and into the world of the workplace, and looked at how the "persuasion grandmasters" — this time those on the *right* side of the law — go about their business. How lawyers, politicians, advertisers, and salespeople gain access to our thought streams — and subtly divert their course.

---

*It should be noted that Luntz has since attempted to distance himself from the Bush administration policy, and now accepts that humans have indeed had a direct impact on global warming. Similarly, it would be interesting to revisit the "drilling-exploration" exercise in the wake of the 2010 BP oil disaster at the Macondo rig in the Gulf of Mexico.

Our survey proved revealing. The brain employs some pretty simple house rules — and if you know how to bend them, persuasion jackpots aren't so hard to hit.

In the following chapter, we continue our tour of the casino of social influence by going "up" a level — from the individual to the group. Suggestion and framing may well take free will to the cleaners, but it's what *others* are doing that often walks off with the pot.

In the days of our ancestors, safety in numbers was vital — and that ancient evolutionary imperative is glacially preserved in our brains.

5

Persuasion by Numbers

An old Irish man is lying on his deathbed, with his son sitting by his side. The old man looks up at him, and says, "Son, it's time for you to get me a Protestant minister."

The son is incredulous. "But, Dad!" he protests. "You've been a staunch Catholic all your life! You're delirious. It's a priest you'll be wanting now, not a minister."

The old man smiles weakly and shakes his head. "Son, please," he says. "It's my last request. Get a minister for me!"

"But, Dad," cries the son, "you're a good Catholic. You've always been a good Catholic. You raised ME as a good Catholic. You don't want a minister at a time like this!"

The old man won't budge.

"Son," he whispers, "if you respect me and love me as a father, you'll go out and get me a Protestant minister right now."

The son relents and does what his father asks of him. Then he and the minister come back to the house, and the minister goes upstairs and converts the old man. As the minister is leaving the house, he passes Father O'Sullivan rushing in through the door. Inwardly delighted, he stares solemnly into the eyes of the priest.

"I'm afraid you're too late, Father," he says. "He's a Protestant now." Father O'Sullivan bounds up the stairs and bursts into the old man's room.

"Seamus! Seamus! Why did you do it?" he cries. "You were such a good Catholic! We went to St. Mary's together! You were there when I performed my first Mass! Why in the world would you do such a thing like this?"

The old man gazes intently at his friend. "Well, Patrick," he says,

"I figured if somebody had to go, it was better one of THEM than one of US."

We allow our ignorance to prevail upon us and make us think we can survive alone, alone in patches, alone in groups, alone in races, even alone in genders.

— MAYA ANGELOU, address to Centenary College
of Louisiana, March 1990

TALL ORDER

I arrived at London City airport to a scene of total chaos. A complete systems failure meant that everyone was being checked in manually and the usually expansive concourse was packed with queues so geometrically double-jointed even Stephen Hawking would have marveled. No one, at this airport, was going anywhere fast. In fact some weren't going anywhere at all.

A guy a little bit ahead of me was spoiling for a fight. He'd already sounded off a couple of times on his phone, and had finally had enough. He marched — or rather, minced — right up to the front of the queue, threw down his Prada cabin bag, and demanded that he be checked in immediately. The assistant was unimpressed.

Slowly getting out of her chair, she climbed, very methodically, on top of her counter — and stood up. Then, in a loud, measured, and exquisitely disdainful tone, she addressed him from on high.

"What makes *you* think that *you* should be treated any differently than anyone else at this airport?"

He never made the flight.

In the last chapter we examined in some detail the dynamics of what we might call cognitive suggestion. How the grandmasters of professional persuasion, the lawyers, the salespeople, the ad men, and the politicians, are able to manipulate not just the kind of information that our brains take in (the raw material of influence), but also precisely what we do with that information once we've got it. But stories like those of the check-in assistant allude to a different kind of influence than that discussed so far — an influence that feeds not so much on the power of information as on the pull of human relations.

Take what happened in Jonestown, for instance. On November 18, 1978, the Reverend Jim Jones tape-recorded a forty-four-minute message instructing over nine hundred members of the People's Temple to drink cyanide-laced Kool-Aid in a remote agricultural community in the jungle of northwestern Guyana. The carnage that followed remains, to this day, the greatest single loss of American civilian life apart from 9/11.*

Then there was London, five years ago. On July 7, 2005, at 8:50 A.M., a primary school teacher, a carpet fitter, and a fish-and-chip shop assistant caused a series of explosions in the bustling city center that obliterated, in cold blood, the lives of thirty-nine people who were on their way to work. Barely one hour later, at 9:47, a fourth member of the death squad — eighteen years old and just out of school — contrived to set off another device, bringing the tally of devastation to fifty-two.

Needless to say, these are extremes. Examples of group influence — of radicalization and brainwashing — so far removed from everyday experience that they appear, to all but a crazed minority, beyond comprehension. Which in many ways they *are*. Yet they, too, bear the thumbprints of persuasion — ancient forces of interpersonal attraction that lash together identities. They comprise a spectrum of influence ranging from simple changes of mind at the one end, to wholesale shifts in worldview at the other. From the trifling affairs of day-to-day subsistence to matters of life and death.

Cast your mind back, for instance, to Solomon Asch and his "line study" in the previous chapter. It was obvious, if you remember, which of those lines were the same length and which were not. But add a few dissenters to the mix — confident, consistent, unanimous dissenters — and things became trickier. Participants began to see the lines not as they were, but as those around them saw them.

And that — when ideologies start getting involved, when the lines turn into dogma — is when it all begins to get dangerous.

CONVICTION BY NUMBERS

Just how easy it is to radicalize a group of moderate — though partisan — individuals is revealed by research into something called *group polarization*.

———

*This statistic refers to nonnatural disasters only.

Group polarization describes what happens to individuals' opinions when they form part of a group. They become more extreme.

This is something you can demonstrate for yourself with the help of a few friends. First, ask them to give you their *individual* opinions — in private — on an issue such as the following:

> An undercover agent operating behind enemy lines gets taken prisoner by opposition forces and is sentenced to twenty years' hard labor in a remote detention center. Conditions at the center are extremely bad, and the chance of rescue minuscule. The agent ponders his predicament — that he will spend the prime of his life in abject, interminable misery — and begins to formulate an escape plan. But if his attempt is foiled and he is recaptured, he will be executed.

Question: If you were advising the agent, what would you consider the acceptable level of risk beyond which an escape attempt should *not* be made? Select from the following options (numbers on the scale represent the probability of being caught: i.e., the chance of capture varies from 10 percent on the left to 90 percent on the right):

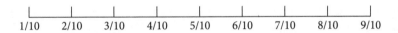

Once your friends have provided their individual opinions, bring them together for Stage 2. This time, you tell them, they must discuss the same issue as a group, at the end of which they must arrive at a joint recommendation.

What you should find is this. If the average of the individual opinions comes out as being *less* than 5/10 (i.e., if it veers toward caution), then the group decision will shift further in that direction (i.e., it will be more conservative than the sum of the individual recommendations). If, on the other hand, the average of the individual opinions comes out as being *greater* than 5/10 (i.e., if it veers toward risk), then the group decision will shift further in *that* direction (i.e., it will be riskier than the sum of the individual recommendations).

The effects of group polarization have been studied in all sorts of settings — from racetracks and shopping malls to the decision-making processes of burglars. In each case, findings exhibit precisely the same pattern.

Venture out as a group and you'll spend more money in the stores. Venture out as a group and you'll burgle . . . fewer houses. (Burglars, when performing collective assessments of a site's vulnerability, tend to be risk averse by nature.)

But it's in relation to prejudice — and, more recently, the rise of extremism — that the most important work has been done. Research has shown that when prejudiced individuals come together to discuss issues of race, their attitudes harden and they become *more* prejudiced. Low-prejudice individuals, on the other hand, become more tolerant (see Figure 5.1).

 Figure 5.1

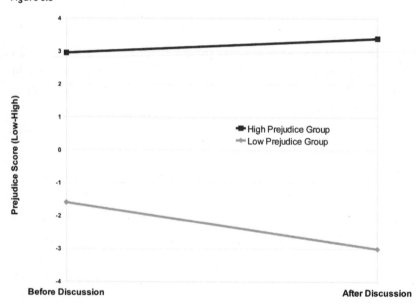

The recruitment tactics employed by many terrorist organizations work along similar lines. The process begins with the targeting of sympathetic individuals (often, at first, through mainstream ideology) who are then brought together in partisan group settings to absorb propaganda and to discuss "the cause."

Twenty-two-year-old Shehzad Tanweer was described by his friends as politically moderate. At school in England he showed promise as a sportsman: cricketer, footballer, and long-distance runner. In 2004 he graduated from Leeds Metropolitan University with a degree in sports science, then

enrolled on a course of "Islamic studies" in Pakistan: in a *madrasa*, in Lahore, with connections — intelligence sources now believe — to an outlawed
Islamist group.

The following July, one sunny summer's morning in London's city center, he blew himself to pieces as the deadly cargo of explosives he'd concealed inside his rucksack blitzed a gaping, crimson hole in one of the major eastern arteries of the Underground.

One sees a similar pattern of transition in the lives of his three associates. Ordinary, regular guys who gradually, through the circles they started
to move in, began to see things "differently."

Who began to see the lines not as they really were. But instead, as others
saw them.

GREEN ROOM

I am, in the interests of simplicity, cutting corners here, of course. In addition to weight of numbers, there's a portfolio of factors that go with increased conformity. These, laboratory studies have shown, include feelings
of incompetence or insecurity; a group presence of at least three (additional
members generate minimal increments in conformity); unanimity (the effect of even a single dissident opinion is catastrophic); admiration for the
group; no prior commitments; and group surveillance of the individual. In
Asch's line study, for example, the incidence of conformity tailed off dramatically when participants, rather than indicating their opinions publicly,
responded in private instead.

Add to these a charismatic leader like Jim Jones, segregation from those
with a different worldview (for members of the People's Temple, dissenting opinion was pretty thin on the ground in the jungle of northwest Guyana — as it was for Shehzad Tanweer in the *madrasa* he visited in Lahore),
and an incremental induction procedure incorporating progressively larger
gestures of group commitment (distributing leaflets, mentoring new members, getting involved in policy decisions: the foot-in-the-door technique,
in other words) and you eventually end up with something very dangerous
indeed. The basic, raw materials of brainwashing. The psychological equivalent of a "dirty bomb."

Yet even with all this there seems to be something missing — a vital
piece of the jigsaw still unaccounted for. Consider, for a moment, the im-

pact — the sheer existential enormity — of a major terrorist attack or mass suicide. Can the events that unfolded in Jonestown, the atrocities of 7/7, the devastation of 9/11 *really* be explained by something as simple as peer pressure? Or are other forces at work? Something deeper, something stronger, something, perhaps, a little more neurological? Are the symptoms of affiliation always as florid as those demonstrated by Asch? Or do strains of the virus sometimes lie dormant — their effects sequestered below the threshold of consciousness?

A clue may lie in the work of Robert Cialdini — fresh from the zoo and that date with the young offenders. In 2007, he and his colleagues conducted a study which tilted at immortality. How, they wanted to know, might hoteliers achieve the impossible: persuade guests — at least once during their stay — to reuse their towels?

Cialdini was interested in the kinds of messages most likely to induce compliance. Would it be those advocating descriptive norms (i.e., those describing how others reuse their towels)? Or the more conventional type of message promoting environmental awareness?

To find out, five cards, each bearing one of the following messages, were randomly assigned to over two hundred hotel rooms, and the number of towels counted up afterward:

- "Help the hotel save energy."
- "Help save the environment."
- "Partner with us to help save the environment."
- "Help save resources for future generations."
- "Join your fellow guests in helping to save the environment (in a study conducted in fall 2003, 75 percent of guests participated in our resource savings program by using their towel more than once)."

Which of the messages do you think was the most effective? Which of them would *you* be most likely to comply with?

If you think it's the last one — "Join your fellow guests in helping to save the environment" — then you're not alone. Forty-four percent of the guests who saw this card in their room reused their towels. The least effective — surprise, surprise — was the one that emphasized the benefit to the hotel: less than 16 percent of guests reusing their towels in this case. And in a follow-up study in which the successful message was nuanced even further to read like this, "Join your fellow guests in helping to save the environment (in a study conducted in fall 2003, 75 percent of the guests who stayed *in this room* participated in our new resource savings program by using their towels more than once)," compliance shot up even higher, to 49 percent.

"If you're in a situation and not sure how to act," comments Noah Goldstein, one of the researchers involved in the study, "you are going to look to other people and the norms of that situation."

Which brings us — rather neatly — straight back to Asch.

Or does it? Let's, for a moment, take a closer look at Cialdini's experiment and compare it with the line study. Notice anything different? Well, to begin with, of course, in Cialdini's experiment there's no departure from verifiable fact. Sure, the majority of guests who stayed in Room 320 might well have recycled their towels. But, hey, it's not like you can get out a ruler and *measure* such behavior — the rightness or wrongness of it — like you can measure the lines in Asch's study.

But that's not the only difference. There's another, more revealing, disparity. In Asch's study, the majority were actually present. They were there, and there was no getting away from them. As a conduit of influence they were, both physically and psychologically, conspicuous. In Cialdini's study, in contrast — well, there *was* no majority. Not one you could see, at any rate. And, more importantly, not one that could see *you*. On paper, sure, they were a force to be reckoned with. But that's in no way the same as being there in the flesh — emerging from behind the shower curtain with a bunch of reusable towels. Yet still they succeeded in persuading the guests to recycle.

The results of Cialdini's study are compelling. There is, they suggest, more to conforming than literally meets the eye. It's not just a case of being seen as the odd one out. Not by a long way. It goes deeper than that. We really do, it would seem, have an inbuilt aversion to going against the grain.

But another body of research goes one step further — suggesting that certain kinds of influence go so deep they can actually affect our *perception*. Quite fundamentally, in fact. And, moreover, that this insidious, mind-bending influence isn't the preserve of the establishment — the great and the good, the guys at the top of the pile — but of another social stratum entirely.

The minority. The underdogs. Those who "see things differently."

SHADES OF INFLUENCE

In 1980, the French social psychologist Serge Moscovici conducted a study that to this day has researchers scratching their heads. The aim of the study was to test-drive Moscovici's "genetic" theory of social influence — that definitive, fundamental change filters upward rather than downward in society. And boy, did it do the trick. Trouble is, no one has managed it since.

Key to Moscovici's theory was a "dual process" model of social influence — the idea that minority influence differs not just quantitatively from that of the majority, but also qualitatively. The minority, Moscovici proposed, works behind closed doors through belief restructuring and cognitive civil war — whereas the majority, as Asch showed, has a different agenda entirely: its persuasive pretensions lying not in getting us to *question* the status quo, but rather to simply *accept* it.

To test the theory was no mean feat. But the paradigm Moscovici came up with sent shock waves through social psychology. And, for that matter, through certain echelons of cognitive psychology, too — where experts speculated on the precise neurological mechanism that might feasibly underpin his conclusions.

At the heart of the experiment lay afterimages — those ghostly hues that float before our eyes whenever we overdose on particular fixes of color. More specifically, negative afterimages — those that comprise a different shade or luster than that of the original stimulus.

Were these images set in stone, as the laws of perception dictated? Or were they — in the right hands — susceptible to influence?

The study may be split into two general phases. In Phase 1, the baseline phase, participants were shown a series of blue slides and were asked, after viewing each one in turn, to write down its color. After visually "detoxing" against a plain white screen, they were also, again after each trial, asked to indicate the color of the afterimage. This they did on a nine-point scale, ranging from yellow/orange at one end (the afterimage of blue) to pink/purple at the other (the afterimage of green).

Once these preliminary measures had been recorded, Moscovici then divided participants up into two groups. One group was told that a fixed proportion of previous volunteers (18.2 percent) had in fact seen the slides as *green* while the remainder (81.8 percent) had seen them as *blue*. The other group was told the opposite: that 81.8 percent had perceived the slides as *green* while the rest of them ... you got it ... saw them as *blue*. Utter nonsense, but hey — enough to install in the minds of those involved a "minority" and "majority" position.

The formalities over, the fun began in earnest. Both groups of participants were then shown another series of slides — fifteen of them this time, all of them the same blue color as those they had seen initially — and were asked, after each one, to specify the color *out loud*. So began the "influence phase" — Phase 2.

But there was a catch.

This time they were joined by an associate who called out, after every trial, GREEN. No nonsense. No two ways about it. GREEN.

And there was more. *This* time, after calling out the color of each of the fifteen slides, participants also had to indicate the color of each of the afterimages — where they fell on the same nine-point scale as before. Did the minority really work in a different way from the majority? Instantiate deeper, lasting, more structural changes in belief: conversion as opposed to compliance? The afterimages held the key.

If Moscovici's theory were to hold water, then the "baseline" afterimages from Phase 1 should, following exposure to minority influence in Phase 2,* shift toward the pink/purple end of the spectrum (the afterimage of green). Consistent, consensual minority dissent, argued Moscovici, made people think. Especially — and this is important — if there were no vested interests. It raised profound questions. *Why* the disagreement? *Why* the departure from the norm? If there's nothing in it for them — and there doesn't appear to be anything — well, there has to be *some* reason for their actions, hasn't there? Somewhere along the line there must be *something*. Maybe they're right. Maybe it's me. Maybe the slide *is* actually green . . .

For those exposed to majority influence, however (in this case, those participants who had been told that 81.8 percent of previous volunteers had seen the slides as green as opposed to 18.2 percent), no such shift in afterimage was predicted. The majority, remember, unlike the minority, just "skimmed the surface." Participants, Moscovici reasoned, might well go along with the majority associate on a superficial level. Declare, *in public*, that — yes — the slides were green. But privately, behind the veil of social self-preservation, it was a different story entirely. In their heart of hearts, or mind of minds, they wouldn't really *believe* what they were saying. Of course they wouldn't. Nothing would really change. The slides would still be blue — just like they were the first time — with the same, cor-

---

*That is, after listening to the judgments of the associate representing the minority of previous volunteers (18.2 percent).

responding afterimage. It would simply be a case of them publicly stating otherwise.

The results of the study, when eventually they came, were incredible. To some, quite literally so. Could there really be a strain of persuasion so virulent as to unweave the fabric of low-level visual perception? Evidently, it seemed that there could.

Just look at the graph in Figure 5.2.

Figure 5.2 — Mean afterimage scores. Higher scores show a shift toward the GREEN afterimage.

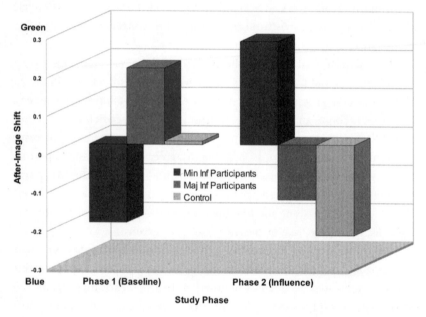

Exactly as Moscovici had predicted, when the associate representing the *minority* position called out green,* the color of the afterimages began creeping toward the purple end of the spectrum, indicative of a real shift in perception. Of hard, subterranean, cognitive rewiring. And this was *despite* the fact that the minority, again as predicted, had little impact on the *public* responses of the participants.

In contrast however, observe what happened in the *majority* group

---

*That is, the associate accompanying those participants told that 18.2 percent of previous volunteers had seen the slides as green.

(those who'd been told that 81.8 percent of previous volunteers had seen the slides as green). Sure, the associate representing the majority position certainly had the beating of their minority counterpart in public, when the participants had to state the color of the slides out loud. But in private, it couldn't have been more different. Here, afterimage perception actually shifted in the opposite direction to purple. Toward yellow/orange.

Minority "withinfluence"? It certainly seemed that way.

TROUBLE ON THE CARDS

Moscovici's findings have proved fiendishly difficult to replicate. But not impossible. Nowadays, in fact, the "dual process" model of group influence is on pretty solid ground — and it's generally accepted that, in contrast to the way conformity works, the minority's job is to "get under the brain's skin." There, provided it remains consistent and is perceived as genuine, it can chisel away at old, established certainties — the things we take for granted — and force us to question the true nature of reality.

At present, we may only speculate. But this — or something very much like it — might well have been what happened to Shehzad Tanweer and his associates in the lead-up to the London bombings. And the followers of the Reverend Jim Jones. Or, upon reflection, something a little more concerted: a dual combination of both minority *and* majority group process.

On one level, it's entirely possible that the effects of minority radicalization really did change the way Tanweer and his accomplices saw the world. Not just metaphorically but neurologically, deep within their brains. On another, it's equally plausible that the pressures of peer group allegiance — of inclusiveness and identity — worked on them in a completely different way: holding them fast, strapping them in, to a trajectory of death they simply couldn't escape.

Here, it wouldn't only have been group forces acting on them either. Once the effects of radicalization had begun to take hold they would have laid themselves open to a number of persuasion viruses — *confirmation bias,* for instance: the tendency that we all have, not just those of us at the outermost reaches of thought reform, to seek out evidence that confirms, rather than disproves, our suppositions. Here's an example:

In Figure 5.3 (overleaf), you have four cards. Each of these cards bears a number on one side and a color patch on the other. As you can see, the sequence at present reads 3, 8, RED, and BROWN. These are the *visible* card faces. But imagine, for a moment, that you're able to pick the cards up — as many of them as you wish — and turn them over.

Figure 5.3 — The Wason four-card selection task.

Question: Which card(s) would you turn over in order to test the truth of the proposition that if a card shows an even number on one side, then its opposite face shows red?

This classic puzzle — the *Wason four-card selection task* — was devised in 1966 by the psychologist, and expert on human reasoning, Peter Cathcart Wason. It is, in actual fact, deceptively simple. Yet this still doesn't prevent nearly everyone who has a go at it from getting it wrong. Yes, I'm afraid it's one of *those.*

Instinctively, most people opt for the 3 and the RED cards. Is that, by any chance, what you went for? If it is, let's just stop and think for a moment about what you were hoping to find.

Let's say, for instance, that you turn over the 3 card to reveal on the back of it RED. Aha, you think, we're in business. But are we? Let's bring in our proposition again and refresh our memories as to its exact wording. The proposition states, "*If* a card shows an even number on one side, *then* its opposite face shows red." Hmmm. Does a 3 and a RED really invalidate this rule? Actually, the answer is no. Just because RED happens to be on the back of the 3 card in no way precludes it from being on the back of a 2.

Likewise, if we flip over the RED card and discover . . . on the back of it . . . 5 — well, this doesn't invalidate the rule either. In fact, we're pretty much in the same boat as before. Just because RED happens to be on the front of a 5 doesn't automatically disqualify it from being on the front of a 4 card.

On the other hand, if we turn over the BROWN card and discover a 4 on the other side of that, then we really *are* in business. This *does* disprove the rule. As does an 8 card with black on the flip side.

So the correct answer, it turns out, is actually 8 and BROWN. It's only by turning over these two cards and attempting to falsify the statement — by actively seeking instances that do not conform to it — that we can challenge its veracity.

But what do most of us do? Most of us — completely unconsciously — seek out instances that *do* conform. We try, most of the time without even realizing it, to confirm what we already know.

SKIN DEEP

This little test provides a concrete demonstration, as well as a gentle reminder, of the power of belief. Of how the stuff we carry around with us inside our heads often forms the selection committee for all further admissions. And it's not as if the test is contentious in any way. No one, after all, has a vested interest in how the cards are going to turn out (if you do, seek help). It's a puzzle, pure and simple.

Back in 1979, psychologists Mark Snyder and Nancy Cantor conducted a now classic experiment that demonstrated the power of confirmation bias not just inside the lab, but outside: when it comes to the kinds of decisions we make on a daily basis. Snyder and Cantor handed participants a description of a person called Jane, which depicted her as being both introverted and extraverted in equal measure. A couple of days later, they got one half to assess her for an extraverted job (estate agent), and the other half to assess her for an introverted job (librarian). What happened? You got it. Each group was better at remembering the attributes best suited to the job they were assessing.

Exactly the same principle lies behind the placebo effect. In an amusing, ingenious (though sadly, unpublished) study which supposedly looked at the influence of subliminal messages on social interaction, a bunch of students had the word SEX daubed on their faces in sunscreen before going out and catching some rays. They were out just long enough for the effect of the sunscreen to become noticeable (to the researcher, that is, not the participants: volunteers were completely unaware of the content of the message) — in other words, for the word SEX to become very lightly emblazoned upon their skin. Then, over the course of the next week, they kept a diary of their social encounters.

Would the subliminal "communication" have any impact on the students' interactions with others?

It sure did. Almost three-quarters of them reported at least one experience that was novel — and which they attributed to the messages on their faces. Such experiences included getting more attention from members of the opposite sex, and better treatment from sales clerks and fellow students.

But here's the deal. In actual fact, only *one-third* of the volunteers had the word SEX etched on their faces. For the rest, it was either a nonsense word written in sunscreen, or a nonsense word written in *water.*

But did that make any difference? Did it heck. The fact that volunteers *believed* there was a hidden message got them looking for things to confirm it. Which, it turned out, were surprisingly easy to find.

BELIEVING IS SEEING

Unfortunately, Jonestown and London 7/7 were not one-offs. On March 26, 1997, thirty-nine members of the Heaven's Gate cult — at the behest of their leader Marshall Applewhite — drank a deadly cocktail of vodka and pheno-barbital (and, for good measure, pulled plastic bags over their heads to finish the job) in order to board the mother ship they believed was descending to Earth. They were found lying neatly in their bunks, all dressed identically in black shirts and sweatpants, brand new black-and-white Nike athletic shoes, and armbands bearing the slogan "Heaven's Gate Away Team." Sadly, they never made it to the game.

It would be easy to laugh — if the consequences hadn't been so tragic. And if the bizarre belief systems didn't have such a terrifying, internal logic behind the warped psychological razor-wire of the group. (This, as we saw earlier, is why cult leaders often set up their communities in remote locations — so as to isolate members from ideological challenge, and cultivate the conditions for what psychologists refer to as *groupthink.*)*

---

*Groupthink, according to Irving Janis, who conducted much of the early work on the phenomenon in the 1970s, comprises "a mode of thinking that people engage in when they are deeply involved in a cohesive in-group, when the members' strivings for unanimity override their motivation to realistically appraise alternative courses of action." The complete inventory of groupthink symptoms runs as follows: feelings of invulnerability creating excessive optimism and encouraging risk taking; discounting of warnings that might challenge assumptions; unquestioned belief in the group's morality, causing members to ignore the consequences of their actions; stereotyped views of enemy leaders; pressure to conform applied to dissenting, "disloyal" group members; shutting down of ideas that deviate from the apparent group consensus; illusion of unanimity; and

But then none of us are immune to confirmation bias. We all do it. Show two opposing football fans an identical tackle and to one it's a foul and the other a fair challenge — depending on the outcome. In fact, given our tribal ancestry, our close-knit ties in the parched primeval badlands of ancient East Africa, it's especially resurgent in those situations in which group affiliation is salient.

Take, for example, the high-profile arrest of Harvard professor Henry Louis Gates at his home in Boston in the summer of 2009 on his return from a trip to China.

What do we know?

Well, Gates, who is black, said that when he got home he found the door jammed — and that he and his driver attempted to force it open. He then said that he gained entry to the house through the back door, and was on the phone to the property's management company when the police arrived.

The police said that Gates became irate after Sergeant James Crowley, who is white, asked him for identification. They say that Gates accused Crowley of being a racist, refused to calm down, and was arrested.

Gates maintains he complied with Crowley's request and turned over identification. He says Crowley arrested him after he'd followed the policeman to the porch, repeatedly demanding the sergeant's name and badge number because he was unhappy with the way he'd been treated.

Crowley has since refused to apologize, saying he followed the accepted protocol.

As should be clear from the discrepancy between these two accounts, one of the parties is being a tad economical with the truth. But which one? Chances are that the side you come down on will have less to do with a detailed examination of the evidence and a hell of a lot more to do with . . . well, what side you're on. If, for example, you buy into institutional racism, or have had your rights as a homeowner infringed in the past, or have been mistreated by the police, then you're likely to see *them* as the culprits. If, on the other hand, you're a hard-line Republican and think that Obama is a Muslim fanatic who's biased toward the terrorists and — whoa, look, here's the proof — he's wading in on behalf of his friend who's black . . . then the

---

"mindguards" — self-appointed members who shield the group from dissenting opinions (Irving L. Janis and Leon Mann, *Decision Making: A Psychological Analysis of Conflict, Choice and Commitment*, New York: Free Press, 1977).

most likely scenario is that Gates had "a chip on his shoulder" and provoked the arresting officer.*

The confirmation bias lies dormant in all of us. Most of us will never join a cult, but all of us are subject to the latent gravitation of our own cherished beliefs. Inside Jonestown, the Reverend Jim Jones's daily salvos confirmed to his followers that their cause was right and that ultimately death would bring about peace and justice. Sound familiar? It should. Switch on your television and listen to reports from Afghanistan.

In fact a recent study by Stanford University psychologists Scott Wiltermuth and Chip Heath suggests that cults and the military have more in common than you may think. Armies train by marching in step. Religions incorporate ritualized singing and chanting into their services.

But why?

Wiltermuth and Heath have discovered that groups whose members engage in synchronous activity tend to be more cohesive — tend to cooperate with each other more — than groups whose members don't. (Even when there are sound financial reasons to cooperate — like the experimenter giving out money!) Might synchrony and ritual therefore have evolved — causing some groups to thrive and others to perish? It's certainly not impossible.

Social psychologist Miles Hewstone asked students of two different faith traditions, Muslim and Hindu, to imagine that a member of their own faith had either helped or ignored them in a time of need. He then got them to do exactly the same for a member of the opposing faith.

Subsequently, the students were asked to guess what might have instigated the behavior of either their Muslim or Hindu counterparts.

Would they, when the opportunity presented itself, come down heavily on the side of their own faith tradition and denigrate the other? Or would they remain evenhanded?

Would they hell. Both Muslims and Hindus cited internal, *personal* factors as the reasons for in-group altruism; and external, *situational* factors for out-group altruism. In other words, in-group members acted of their own volition and of their own good natures — and, what's more, would act

---

*After the incident went national, Obama ended up inviting both of the protagonists — Gates and Crowley — to join him for a beer at the White House, at what quickly became known as the "beer summit." Though neither party apologized for their roles in the affair, they agreed to disagree and promised to have further dialogue.

in precisely the same way if the same thing happened again — whereas out-group members were perceived as having little choice in the matter, and, in the event of an action replay, as being unlikely to repeat such behavior.

In contrast, however, when it came to the question of members of the in-group *not* helping, there ensued, on both sides, an uncomfortable period of throat clearing. Just as had been the case in the "surprising" event of the out-group helping, indifference was chalked up to the situation. Hands were tied. It was an unfortunate one-off. And as for the *out-group* not helping? Well that was easy, wasn't it? It was them all over. Inconsiderate. Unprincipled. And addled with self-interest.

And it's not just what we believe about *others* that influences how we see things. Equally important is how we perceive *ourselves*. In the 2006 soccer World Cup in Germany, German police fêted English fans — not exactly noted for their temperance on such occasions — as being the "best fans in the world." The tournament passed off without incident.

Not, of course, that the compliment was genuine. You must be joking. Rather, the Germans had done their homework. Research has shown that providing individuals with false feedback about themselves can actually induce them to confirm it: to behave in a manner consistent with that feedback. They become the person that they *believe* themselves to be. Or, more accurately, the person that they believe *others* believe them to be. Which in theory, of course, can be anything.

ALLIANCE DEN

In August 2006, an elderly resident of the Strasshof district of northeastern Vienna picked up her telephone and dialed 911. A distressed and disheveled young woman had hammered on her kitchen window begging her to call the police. Some minutes later a squad car pulled up outside. A row with her boyfriend, the frayed ends of an all-night party — there could have been any number of routine explanations for the call. But not in this case. The woman in question turned out to be Natascha Kampusch. And her story, it emerged, was anything but routine.

Eight years previously, aged just ten, Natascha Kampusch had vanished into thin air on her way to school. Her disappearance, at the time, had been all over the Austrian media — front-page news for at least a couple of weeks — and a nationwide search had ensued. There were divers and dogs, a dedicated police unit and civilian volunteers. Even the Hungarians got involved. But all of it had come to nothing. Until now.

In fact, for the whole time she'd been missing, Natascha Kampusch had been literally right there under their noses. In a scene that could easily have sprung from the pages of a Stephen King novel, she'd spent the greater part of the intervening years imprisoned in a dungeon she'd believed to be rigged with explosives.

Alone.

Throughout her ordeal, during the entire period of her extraordinary subterranean imprisonment, her only means of human interaction had been with her abductor, thirty-six-year-old communications technician Wolfgang Priklopil. It was he who'd effectively brought her up, providing her with food, clothes — everything a ten-year-old could wish for. Everything an eighteen-year-old could wish for. Except freedom. *That,* unfortunately, was where Priklopil drew the line.

"He gave her books, even taught her how to read and how to write," reported one of the investigators in the case. "And mathematics and all things like this, according to what she told us."

The dungeon measured just 4 meters by 3 meters, and had a door 50 centimeters by 50 centimeters.

Completely soundproofed, it was sealed in an underground garage.

And, like Natascha Kampusch herself, would probably never have come to light had she not made a bid for freedom while hoovering her captor's car.*

Miles Hewstone's study with the Muslim and Hindu students shows what can happen when group identity suddenly becomes salient. We deify those who are like us, and vilify those who aren't. We believe what we want to believe. But not all intergroup dynamics work this way round. Under certain, exceptional, circumstances we find ourselves believing what we *don't* want to believe. And helping, even liking, those who do us harm.

Take a condition known as the *Stockholm syndrome* — a phenomenon well documented in the literature on hostage negotiation, and perhaps even better documented in the mind of Natascha Kampusch.

The Stockholm syndrome refers to a psychological dynamic in which

---

*Kampusch *was* allowed out of the dungeon (located beneath Priklopil's house) for limited periods to assist her captor with chores. This was on the understanding that if she attempted to escape he would kill her.

hostages come round to liking, even supporting, their captors. Typically this follows conciliatory gestures on the part of the captors, which run counter to the expectation of the hostages. Such gestures might begin with something as simple as the making of a cup of tea or the sharing of a bar of chocolate — and extend right the way along to requests for medical assistance or help "on the outside." Even, in some instances, appeals for emotional support.

And then there are the *really* extreme cases — like that of Natascha Kampusch. Here, the exploits of her captor Wolfgang Priklopil didn't just stop at tea. Nor did they stop at chocolate. Instead, they ran the whole gamut of a father-daughter relationship, from the provision of food and clothing to that of a full-time education. And, what's more, not just for a period of a few days, but for eight years. Just think, for a moment, of the level of emotional dissonance such intensity of commitment would foment; the saturnine forces of mind that must have been pulsing backward and forward within the cramped and loaded compass of that dungeon. Are we really surprised, even under such appalling confinement as this, that *some* degree of captor-captive bonding might have surfaced?

Precisely how the Stockholm syndrome works is complex. It acts, for the most part, through a double hit of reciprocity and consistency — that lethal cocktail of influence we encountered in the previous chapter courtesy of telesales employee Pat Reynolds. The fulcrum of the dynamic is the power differential between captor and captive. Conciliatory behavior by the captor sets up an imbalance in the mind of the captive between their *feelings* toward the captor (negative) and the *actions* of the captor (positive). Powerless to change the actions of the captor, the captive has only one means available to her — noxious though it may be — by which to restore cognitive consistency: change her attitude toward such actions. Add to this our old friend the reciprocity principle — altruistic gestures should be paid back in kind — and the results, as we have seen, can be devastating.

But reciprocity and consistency aren't the only culprits here. As Marshall Applewhite, Jim Jones, and others like them know only too well, one of the biggest secrets of mind control is control over everything else.

RUFF WITH THE SMOOTH

In the mid-1960s, cognitive psychologist Martin Seligman stumbled, partially by accident, on a rather curious phenomenon. It began with a routine conditioning experiment. Dogs, in line with the usual conditioning protocol, were exposed to a pair of stimuli in quick succession — a tone, followed by a harmless, though painful, electric shock — the aim being, through repeated association between the two, to elicit fear just of the tone itself.

In order to ensure that the preliminary association between tone and shock was properly established, Seligman restrained the dogs in the initial conditioning phase of the study so that, following the onset of the tone, exposure to shock was inevitable. They couldn't, in other words, get away. But during the "test phase" — in which the tone appeared on its own — things were different. The dogs had the chance to escape — evidence, were they to take it, that the conditioning had proved successful.

The experiment went badly wrong. And wrong in a way that no one could have predicted. To Seligman's amazement . . . nothing happened. Nothing whatsoever. Even though, in the test phase, the dogs had a clear escape route whenever the tone sounded, they simply stayed put. Incredibly, they made no attempt whatsoever to evade the "impending" shock.

Even more incredible was what happened next — when Seligman dispensed with the tones altogether and just administered shocks. Real ones. The dogs *still* didn't move. Feelings of *learned helplessness* — the term coined by Seligman to describe such behavior — had hijacked the animals' brains and taken their "reasoning" hostage. So much so that they simply no longer cared.

Today, Martin Seligman is still making waves. In 2002, in San Diego, he popped up at a forum organized by the CIA as part of the U.S. military's SERE (survival, evasion, resistance, escape) program — a course specifically designed to inoculate pilots, Special Forces personnel, and other high-value captives against torture. Or, if you prefer your definitions unabridged, techniques of interrogation explicitly outlawed by the Geneva Convention. There, to an audience of psychologists and other U.S. government officials, Seligman expatiated for three hours on — yes, you've guessed it — the dynamics of learned helplessness. Though he has since repudiated — and repudiated quite robustly — even the merest hint of a suggestion that he might willfully have been associated with the formulation of so-called torture programs, those present at the forum included a number of key U.S. military personnel who later proved instrumental in developing techniques of "enhanced interrogation."

Of course, some people are more prone to feelings of learned helplessness than others. It depends on your *attributional style* — or, to put it another way, the way you think about the things that happen to you in life.

Both positive and negative outcomes may be perceived as a function of two psychological dimensions:

1. Locus of control — whether you infer an internal cause for the outcome and take personal responsibility for it versus whether you infer an external cause and put it down to the situation (examples of both of which we saw in Miles Hewstone's Muslim/Hindu study);
2. Generality — whether you see the outcome as a specific one-off or as something longer term.

Imagine, for example, that you've just failed an exam. On the basis of these two dimensions, there are four different ways you can rationalize your performance:

LOCUS

GENERALITY	Internal	External
Specific	I didn't work hard enough	This exam wasn't a true test of my ability
General	I'm never any good at exams	Exams in general aren't a true reflection on ability

If you're a pessimist, or prone to depression, then for *negative* outcomes like this you're more likely to have a *general/internal* attributional style (bottom left box) — and be at greater risk of learned helplessness than someone who sees things more *specifically*.

In contrast, however, now imagine the following. You've just been handed your quarterly report by your broker and discovered that shares in a new stock you bought have gone through the roof. Again, according to the two dimensions, there are four different ways you can look at the situation:

LOCUS

GENERALITY	Internal	External
Specific	I got lucky and hit the jackpot this time round	The company was well managed during this quarter
General	I'm generally a pretty good judge of the market	The economy is in good shape — I made hay while the sun was shining

Here, when the outcome is *positive,* attributional styles reverse. It's the *optimist* who displays a general/internal profile (bottom left) — while the *pessimist* is more specific.

In short, optimists take the credit for good outcomes and contextualize bad ones, while pessimists do the opposite: externalize the good times and blame the bad things on themselves.*

But here's the deal. Manipulate someone's environment for long enough — flood him with stimuli he continually has no control over — and sooner or later attributions will start to change. Just like the dogs in Seligman's experiment, external metastasizes into internal — and cancer of the will develops. A study in the 1970s played tape recordings of office machinery as volunteers solved puzzles. Guess what? They performed better when they believed the noise to be controllable than when they believed it to be uncontrollable: even though it was exactly the same noise played at exactly the same volume.

Even in routine police work, in interview rooms and custody suites a million miles away from the austerity of military detention centers, the dynamics of control play a crucial role in obtaining information. Especially, it would seem, when it's in otherwise short supply.

One senior British detective told me:

> Think about it. Some of the people we get in here are used to being top dog. To getting their own way. We have gang leaders, wife beaters, you name it. But as soon as you come through those doors, the roles are reversed. We control everything that happens to you in here. Every move you make. Anything you want — it's entirely up to us. It's up to us when you can have a cup of tea. It's up to us when you can take a piss. It's up to us whether the lights are turned on or off in your cell. All those little things you tend to take for granted when you're at home — forget it.
>
> Soon as you get in here, *we're* in charge. We can look in on you anytime we feel like it through an observation flap in the door. And if we don't want to talk to you we can just snap it shut. See what I mean? When I say we control everything, I mean *everything.* And they're just not used to that, a lot of the people we get in here. They don't like it one fucking bit when the boot's on the other foot. But sooner or later most of them start to get it.

---

*To find out what your attributional style is, why not complete the questionnaire at the end of the chapter?

STICKING POINT

One sees exactly the same in cults. Alongside those factors we looked at earlier — those designed to increase the flow of conformity — cult leaders follow a pattern: a curriculum of influence as devastating as it is predictable. Jonestown was situated in the wilds of northwestern Guyana, where the travail of "getting out" in many cases outweighed the benefits (links with external friends and family having gradually eroded over time). Jones's voice droned twenty-four hours a day from an intercom system — not so much washing the brain as waterlogging it — and the children of followers were encouraged to call him Daddy. Slowly, insidiously, systematically, Jones — through monotonous, consistent persistence — willed himself into God. First he was every*where*. Then he was every*thing*.

Victims of domestic violence face an almost identical scenario. Listen to Lisa, a thirty-five-year-old mother of two:

> It started with my friends. He'd say, "You're too good for her!" and then gradually that would be that. Whoever it was I slowly lost contact with. Same thing happened with my family. He said my mum was against him, my brother was against him, so why did I have anything to do with them? Even meeting up for a cup of tea was seen as taking sides. He'd drop me to work at nine and pick me up again at five so that I wouldn't have time to socialize with anyone. And he'd ring me up at lunchtime to see if I was alone. As for money, I didn't so much as sniff my salary for nearly a year and a half — he'd arranged for it to be paid straight into his account. . . .
>
> The violence kicked off with my clothes. If we were going out anywhere and I decided to dress up and wear makeup, he'd hit me and call me a slut. And if I didn't dress up, he'd hit me for not making an effort. I couldn't win. Toward the end, he was even examining my underwear to see if I'd had sex with anyone. That, really, was what did it for me. That was the final straw.

Cases like Lisa's seem unbelievable when you set them out cold on the printed page like this. Yet ask the domestic abuse team of any police force in the country whether the details stack up and they'll tell you exactly the same thing: they come across hundreds of such incidences a year.

Police Constable Andy Green of the Cambridgeshire Constabulary talks me through the profiles of the different types of offender. As he does so, I'm suddenly struck by the ubiquity of the list: these descriptions, as well as relating to *domestic* abuse, could be applied just as easily to the *workplace*. There is, from my own personal viewpoint, at least one former colleague I recognize in there!

Green nods in agreement. "Absolutely," he says. "These, basically, are just styles of persuasion that may be put into practice anywhere. Just because they've been identified as being present inside the home doesn't mean to say that they can't be found anywhere else in life. In other contexts. They're just different means to the same unfortunate end."

The taxonomy that Andy Green runs by me might best be described as "semiofficial." It hasn't been ratified as such, but is based on years of operational experience and has made it into a booklet. The inventory ranges from the Bully who shouts and sulks, to the Head Worker who puts you down — telling you you're ugly, or stupid, or useless. Or all three combined. Others include the King of the Castle, who treats you as a servant, the Liar — "Loosen up, it's only a bit of fun" — and the Persuader, who threatens, praises, and flatters in equal measure.

"Often," Green adds, "the effects of the manipulation are so strong that even when you open the door for [the victim] and say, 'Here, you can leave . . . there's a place you can go . . . don't let him do this to you anymore' . . . they look at you as if you're crazy. 'It'll just make him angry,' they say. Or, 'He doesn't really mean it.' It's like their brains have been immobilized by month upon month, year upon year, of being told the same thing over and over. Like they've been infected by some kind of virus."

When I tell him about Martin Seligman, Green shakes his head.

"I wish I could say that was news," he says. "But I can't."

Once, several years ago now, at a workshop on autosuggestion, I myself was infected by the immobilization virus at the hands of an ex–Special Forces martial arts instructor. I forget his name now, but let's call him Curt. Curt began the workshop by lining ten of us up against a wall and telling us to clasp our hands together as tightly as possible. He then told us that over the space of the next couple of minutes or so he was going to hack into our brains and hijack our free will. Quietly. Covertly. But mercilessly. Meanwhile, he said, we were to continue clasping our hands.

Curt's assertion was greeted with suspicion. Though not, I must confess (and I'm sure I wasn't the only one), without some degree of foreboding. I already knew a little bit about Special Forces — the kinds of things they were capable of. Could this be some kind of trick? Had Curt somehow managed to dab glue on our hands without us noticing? To be honest, I wasn't certain.

Sure enough, over the next couple of minutes, Curt got to work. "You'll start to feel your hands slowly getting stuck together," he intoned, "as if they're being held in place by a very powerful adhesive. As you feel this," he added, "you are to press them together even tighter, to facilitate the bonding process and to make it as strong as possible. Finger by finger," he continued — in a relaxed, deliberative, but completely authoritative voice — "you are to cement your hands into position so that, even if you wanted to, you wouldn't be able to move them."

He went along each one of us in turn.

"Make the bond absolutely rock solid," he said, cupping his hands around ours and increasing the pressure even more. "In fact," he said, "make it *so* solid that nothing, absolutely nothing, will be able to prise it apart."

Curt went on like this for another minute or so — confidently, methodically, and matter-of-factly encouraging us to cement our hands together. This is insane, I thought to myself, as I clenched my fingers as hard as I possibly could.

Then suddenly I started to panic.

What if the joke was on us? I thought. And our hands really were stuck together? What then? Was he going to mug us? Maybe the whole autosuggestion gig was some elaborate low-level scam specifically designed to get a load of suggestible schmucks like us together in a room forking out a hefty enrollment fee. And look — everything was working like clockwork! Maybe, once we'd written out our checks, Curt was going to skim off what was left. From our credit cards. While we were all stuck together with Superglue.

The conniving bastard, I thought.

That was *it*, wasn't it? Of course it was. How could I have been so stupid? It was *us* who were crazy, not him.

Calmly, frantically, I started ruminating. My wallet . . . how much was in it . . . hmmm, don't know. . . . Canceling the cards would be a pain in the arse . . . but hey . . . better than getting SHOT. . . . What about photo IDs . . . well, for a start there was my driving license. . . .

Meanwhile, I kept clasping my fingers.

Until, suddenly, Curt just stopped.

"Right," he said. "What I want you to do now is stop pressing your hands together and slowly release your fingers. Do this on the count of three. Are you ready? One . . . two . . . three."

We all eyed each other uneasily. I glanced at the guy standing next to me, and he glanced back. "I'm not sure about this," he mouthed. "Me neither," I mouthed back. I realized I was sweating. Then we started unclasping. Some people managed it instantly. And immediately reached for their back pockets. Others, like myself, found it more difficult. But one or two simply found it impossible. Their hands really *were* stuck fast! Just as Curt had predicted, try as they might, they just couldn't prise them apart.

Eventually, of course, when the dust had settled and things had died down a bit, they *did* succeed in extricating themselves. Then we all shook our heads and laughed. Ho-ho.

But the lesson, as any good stage magician knows, was clear as day. Tell someone something often enough — and some of them, at some point, will come round to believing you.

Believing you, no matter what.

SUMMARY

In this chapter, we've seen how an ancient ancestral force field buried deep within the brain — the need to conform — can wield just as much influence over our attitudes and behavior as any of the persuasion strategies employed by modern-day advertisers and opinion formers. Old habits die hard, and the actions of those around us — especially those similar to ourselves — place powerful evolutionary magnets by the side of our brains' belief compasses. Conformity, indelibly, is written into our genes. During the time of our forebears, when "survival" and "group" were more or less synonymous, the market for individuality was somewhat less buoyant than it is today — and the ability to "keep one's head down" would almost certainly have conferred an advantage. It's a lesson we've never forgotten.

In a world fueled by competing ideologies, our tribal origins can sometimes give cause for concern. Group dynamics conform to certain laws, and those with a knowledge of such laws can, if so inclined, "genetically modify" a group to create, within society, mutant strains of extremism far removed from the norm. But not all groups follow the *same* laws. And while the power of greater numbers reforms us from "on high," the minority works "from within," nudging the brain into questioning reality — into unpicking, then restoring, the transformative fabric of truth.

In the following chapter, we turn the spotlight fully on split-second persuasion — placing it under the microscope and mapping its DNA.

Is there, we ask, concealed within the melody of mind, a golden chord of influence that *all* of us can play? Not just the persuasion virtuosi, but the street performers, too?

The answer, it turns out, is yes. Our analysis uncovers the double helix of influence, enshrined within which lies persuasion's secret code.

Attributional Style Test

The following ten statements refer to different ways of looking at life events. Indicate on the scale provided the extent to which you either agree or disagree with each one.

For example, if you strongly agree with the statement, circle 4. If you strongly disagree, circle 1. The scale will appear at the end of each statement.

1. When I perform well on a task at work or sail through an exam it's mainly because it was easy.

    1   2   3   4

    Strongly disagree   *Strongly agree*

2. If I fail an exam I can do better next time by studying harder.

    1   2   3   4

    Strongly disagree   *Strongly agree*

3. "Right place, right time" is the recipe for success.

    1   2   3   4

    Strongly disagree   *Strongly agree*

4. Attending political rallies is usually ineffective: nobody takes much notice.

    1   2   3   4

    Strongly disagree   *Strongly agree*

5. Intelligence is determined at birth — there's not much you can do about it.

    1   2   3   4

    Strongly disagree   *Strongly agree*

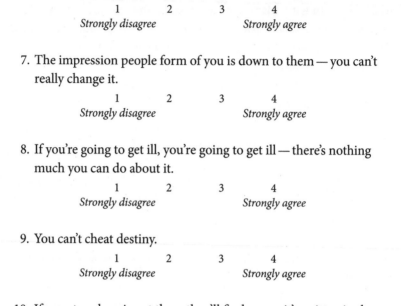

6. I attribute my successes to my abilities rather than to chance.

    1   2   3   4
    Strongly disagree   *Strongly agree*

7. The impression people form of you is down to them — you can't really change it.

    1   2   3   4
    Strongly disagree   *Strongly agree*

8. If you're going to get ill, you're going to get ill — there's nothing much you can do about it.

    1   2   3   4
    Strongly disagree   *Strongly agree*

9. You can't cheat destiny.

    1   2   3   4
    Strongly disagree   *Strongly agree*

10. If your true love is out there they'll find you — it's written in the stars.

    1   2   3   4
    Strongly disagree   *Strongly agree*

SCORING: For Items 2 and 6, reverse your score so that 1 = 4 and 2 = 3 and so on. Then total your score for all ten items. Scores of 15 or below generally indicate an *internal* attributional style, whereas scores above 25 indicate an *external* attributional style. Scores in the 15 to 25 range indicate a mixture of both.

6

Split-Second Persuasion

A London to Cape Town flight runs into severe turbulence over the jungles of central Africa. Word that some of the passengers are extremely nervous reaches the cockpit. A few moments later, the pilot's voice is heard.

"Jesus, we're all going to die! We're all going to die!" he screams. "Oh, shit! That was the intercom light not the engine light."

The plane erupts with laughter and calm is restored.*

Graham Chapman, coauthor of the parrot sketch, is no more. He has ceased to be. Bereft of life, he rests in peace. He's kicked the bucket, hopped the twig, bit the dust, snuffed it, breathed his last and gone to meet the great Head of Light Entertainment in the sky. And I guess that we're all thinking how sad it is that a man of such talent, of such capability for kindness, of such unusual intelligence, should now so suddenly be spirited away at the age of only forty-eight before he'd achieved many of the things of which he was capable and before he'd had enough fun.

Well, I feel that I should say, "Nonsense. Good riddance to him, the freeloading bastard, I hope he fries!" And the reason I feel I should say this, is he would never forgive me if I didn't. If I threw away this glorious opportunity to shock you all on his behalf. Anything for him but mindless good taste.

I can hear him whispering in my ear last night, as I was writing this. "All right, Cleese," he was saying, "you're very proud of being the very first person ever to say shit on British television. If this service

---

*Aficionados of Gary Larson's *Far Side* may have come across a similar line in their favorite cartoon series. The captain of this particular plane was evidently a fan.

is really for me, just for starters I want you to become the first person
ever at a British memorial service to say fuck."

— JOHN CLEESE, oration at Graham Chapman's funeral, 1989

PERSUASION GENIUS

One afternoon, in a classroom in rural Germany, a teacher sets his students
the following problem. Add up, he says, all the numbers between one and a
hundred. He goes to the blackboard and sketches out the sum:

$1 + 2 + 3 \ldots 98 + 99 + 100$.

And then he sits down and takes out a pile of paperwork.

The kids in his class are only seven years old. So the teacher presumes
it'll take them the rest of the day. Just what he needs to finish off his chores.
But then, after twenty seconds or so, one of them puts up his hand.

"Sir," says the boy, "I think I've got the answer." "Nonsense!" says the
teacher. "It's 5,050," says the boy. The teacher is dumbstruck. He approaches
the boy and asks him to explain himself. How did he manage to find the solution so quickly?

"It's simple," says the boy.

He goes to the blackboard and starts writing:

$100 + 1 = 101$

$99 + 2 = 101$

$98 + 3 = 101$

Then, suddenly, he stops.

"See," he says. "There's a pattern. Between one and a hundred there are
fifty pairs of numbers that add up to one hundred and one. So the answer
must be fifty multiplied by one hundred and one. Which is five thousand
and fifty."

Some years later, among a host of other discoveries, Carl Friedrich
Gauss developed modular arithmetic — a major contribution to the field of
number theory — and is recognized today as one of the greatest mathemati-
cians in history.

I love this story about Carl Friedrich Gauss. Whether it's actually true
or not, I have no idea. But that's not really the point. What I like about it is
the math. Its algorithmic secrecy. I like the idea that embedded within the
tedious sequence of numbers is a clean and simple pattern. A pattern that
reveals, were we only able to discern it, a neat and elegant solution.

What's true of math is also true of persuasion. Faced with a problem that needs working out, most of us go the long way round and do what we've learned in the classroom: add up the numbers. But then we have the geniuses. The people who don't just hit the nail on the head, they kick it in the nuts for good measure. Imagine you've been asked to give a eulogy at the funeral of one of your best friends. You take your place in front of the congregation and start to go through the motions. 1 + 2 + 3 . . .

"He was a great friend and will be sadly missed. Blah, blah, blah . . ."

Which is fine. You get there in the end.

But *now* let's imagine you solve things slightly differently: "Graham Chapman, coauthor of the parrot sketch, is no more . . ."*

5,050.

Alternatively, imagine that you're the captain of a plane flying through severe turbulence and that your passengers are terrified. What do you do? Well, you could explain to them that air travel is actually one of the safest forms of transport. That turbulence isn't dangerous. And that you'll soon be through the worst of it . . . 1 + 2 + 3 . . . Or you could do what the pilot did on that London to Cape Town flight. Settle the nerves with a single knockout sentence.

Finally, put yourself in the position of career cop Ron Cooper: twenty-three years in the job and you're faced with a man a hundred feet up. It's up to you to talk him down. You pull out your calculator and start punching in the numbers.

"Why don't you just take a couple of steps back for a moment, I'm sure we can work this out."

Or do you?

"Mind if I take my jacket off?" Cooper asks. "You get a bit hot running up fourteen flights of stairs."

"Do what you want," says the guy. "I don't give a fuck."

Slowly and painstakingly, in howling wind and pouring rain, Cooper starts undoing the buttons of his police overcoat. Twenty minutes — and fourteen stories — earlier, he'd been first on the scene when the call had come through. Young man. Aged around twenty-five. On roof of multistory car park. Threatening to jump.

---

*This has to be seen to be believed. Check out Cleese's tribute to his former colleague on YouTube: Graham Chapman's Funeral.

"The world's a pile of shit!" the guy had shouted to the gathering crowd of onlookers down below. "No one gives a fuck anymore. No one cares whether I live or die. So why should it bother *me*?"

Cooper takes off his coat. Then his tie. And then, with the guy on the ledge eyeballing him all the time, starts to unbutton his shirt.

"Don't try anything funny," the guy says, as Cooper is about to remove it. "Or I'm off!"

"'Course not," says Cooper as he folds it up neatly and puts it to one side. "Just trying to get comfortable, that's all."

He's down to just a T-shirt as the wind continues to howl and the rain turns to sleet.

PISS OFF — I'VE GOT ENOUGH FRIENDS! reads the slogan on the front.

He maneuvers himself onto the ledge and then turns to face the young man head on, so the slogan is fully visible. He looks him in the eye.

"Right then," he says. "You want to talk about it or what?"

ANATOMY OF INFLUENCE

The solutions that Ron Cooper, John Cleese, and the airline pilot came up with to their very different predicaments worked brilliantly. (You'll be pleased to know that the guy on the ledge saw the funny side of Cooper's T-shirt.) But no two people think alike. Their solutions worked for *them*, at that particular moment. And lucky for them that they did.

Such an observation has important implications for the way we've been looking at persuasion up until now. The funeral speech. The talking down. The nervous passengers. There could, in theory, be any number of solutions to such problems. Equally irreverent (or not, as the case may be). And equally "Gaussian." It depends on who you are. And who, even more importantly perhaps, your audience may be.

On the other hand, however, we've also been looking at a system. A formula. A persuasion algorithm — that, if correct, would seem to factorize such variation, such diversity of style, into a triad of rhetorical constants:

1. The basic raw material of what you say — what your audience pays *attention* to
2. The manner in which you deliver that raw material — a major predictor of how your audience will process or *approach* it

3. Psychosocial factors relating to how your audience evaluates you, or appraises what you say, in the context of their relationship with others: the parameters of *affiliation*

So how do we reconcile these two contrasting standpoints? Is *all* successful persuasion, in its many different guises, subsumed within the Three A's? Or is there something else: something Keith Barrett's been missing down the years?

To answer this question I became a persuasion collector. Over a period of eighteen months, from a wide variety of sources, I put together an "influence bank": a definitive anthology, of over 150 examples, of sudden, dramatic turnarounds. Like the one enacted by Ron Cooper. Or, if you remember back to the Introduction, by the musician on the airplane. Examples such as these — of split-second persuasion, as I called it — were integral, I figured, to mapping the genome of influence. If the Three A's really were endemic to persuasion, if they really did constitute the basic building blocks of mind control, then where would they most likely show up? In the game changers, and cliffhangers, and matters of life and death? Or over Michelin-starred profiteroles in first class (assuming, of course, there wasn't a total arsehole in the cabin)?

Once the database had started to gather momentum, I found some volunteers. Read over these scenarios, I told them. And then, for each one, write down the factors that you think most contributed to the persuasive outcome. The results were astonishing. Analysis revealed there were five major axes of persuasion:

1. Simplicity
2. Perceived self-interest
3. Incongruity
4. Confidence
5. Empathy

Or SPICE for short.

Remarkably, these five factors fitted perfectly not just with Keith Barrett's Three A's (simplicity and incongruity mapping onto attention; confidence onto approach; and perceived self-interest and empathy onto affiliation) but also incorporated those principles we saw earlier were so integral to influence in the animal kingdom. As well, of course, as to neonatal persuasion.

Here was an influence that united all influences. That had an incubation period of seconds. A strain of persuasion so immediate, so dangerous, so *ancient* — it didn't just turn the tables, it kicked 'em over. In the number of words (just ask Ron Cooper) you could slap on the front of a T-shirt.

SIMPLICITY

COLOR BLIND

A few years ago now, I remember a local London newspaper carrying the story of an elderly Afro-Caribbean man traveling home from work on a bus. At one of the stops a drunk guy got on and couldn't find a seat.

"Get up, you fat black nigger bastard!" he shouted at the man.

"You calling me fat?" said the man.

The bus erupted with laughter, and the drunk guy got off.

Disaster averted in just four amazing words.

The golden rule of any kind of persuasion — from politics to advertising, from negotiating that contract to preventing a large-scale race riot — is that it's not what you say that necessarily wins the day, but how you say it. Often, the simpler the better.

Research has indicated time and time again that our brains have a bias for simplicity. Consider, for example, the addition problem below. Cover it up with a piece of paper then work your way down to reveal each number in sequence as you add them up in your head:

$$
\begin{array}{r}
1,000 \\
40 \\
1,000 \\
30 \\
1,000 \\
20 \\
1,000 \\
+\ 10 \\
\hline
\end{array}
$$

What answer did you get? If you came up with 5,000 try again. In fact the correct answer is 4,100. So what went wrong? Well, when the brain reaches the penultimate subtotal of 4,090 it *expects* that the final total will be a nice and cuddly round number. So it takes a gamble on the one that comes quickest and easiest to mind: 5,000.

FIRST CLASS INFLUENCE

The sense of fluency that the brain gets when it processes information is a key predictor of whether it's going to "run with it" or not. Simple is good. Complicated is bad. This is why split-second persuasion is so powerful. It comprises, in zoological terms, the modern-day human equivalent of a key stimulus of influence. Like the world's top exponents of the martial arts, some of whom, as we shall see later, are well into their eighties, the split-second persuader wastes little energy. In exactly the same way as these shadowy monks and ghostly grandmasters zero in on *physical* pressure points, the split-second persuader, in turn, goes straight for the psychological jugular.

In splt-scnd prsuasn, in othr wrds, only infrmtn essntl 2 d commnctn of d mssge is inclded in d mssge.

Luke Conway, professor of psychology at the University of Montana, studies the role of simplicity in political oratory. What he's turned up is intriguing. Conway has found that when politicians are running for election . . . guess what? They go back to basics and their policies get less fancy.

Conway analyzed the four State of the Union addresses of forty-one American presidents in their first terms, and detected a pattern. He discovered that the longer a president served in office, the greater the decline in ideological complexity. The correlation was linear. Inaugural State of the Union addresses were nuanced. They were inclusive in tone — typically more embracing of opposing points of view. And, conceptually speaking, more fussy. The last address — the one immediately preceding potential re-election — was, for a politician, about as nailed down as you can get.

"Simplicity sells," concludes Conway. "No one marches to rallying cries that say, 'I may be right, I may be wrong, let's dialogue.'"

Take, for instance, one of the greatest political clarion calls in history. When Winston Churchill delivered his immortal "We shall fight them on the beaches . . ." speech on June 4, 1940, in the wake of the withdrawal of the British Expeditionary Force from Dunkirk, he could have put it differently. Instead of going down as one of the greatest pieces of oratory ever spoken, Churchill's address to his people might have gone something like this: "Hostilities will be engaged with our adversary on the coastal perimeter ..."

Unfortunately, we shall never know why Churchill settled on the particular version he did. People do funny things under pressure, don't they? But, on a slightly less dramatic note, we *do* know why the marketing department of Marks & Spencer have recently put "Exclusively for Everyone" on the sides of their transportation fleet.

"There's not much room on the side of a truck," a spokesman told me when I called. "Even less on a van. And if you're moving in traffic, not much time to read it. I suppose we *could* have put, 'We have top-quality merchandise which is affordable and widely available.' But somehow it doesn't have the same ring to it, does it? In advertising, it's best to keep things simple."

Matthew McGlone, at the University of Texas in Austin, and coworker Jessica Tofighbakhsh have conducted an experiment on poetry. Well, not poetry exactly. More rhyme. In an ingenious take on the anatomy of wisdom, McGlone and Tofighbakhsh set out to investigate whether statements that rhyme contain more truth — or rather, are *perceived* as containing more truth — than those that don't.

To get the ball rolling, McGlone and Tofighbakhsh assembled a portfolio of pithy, but obscure, aphorisms. Next they added their own, somewhat discordant, modifications to the mix. "Caution and measure will win you treasure" they amended to "Caution and measure will win you riches." And "What sobriety conceals, alcohol reveals" to "What sobriety conceals, alcohol unmasks." They then got a group of volunteers to read over their stockpile of wisdom — both the original and the modified offerings — and rate each one for accuracy. How well, McGlone and Tofighbakhsh asked them, did the proverbs stack up against real life?

Just as we might have predicted — and just as McGlone and Tofighbakhsh *did* predict — they went for resonance. Participants perceived the original statements that rhymed as being less cutesy and more genuine than the modified ones that didn't. As providing a truer, more accurate reflection of the way things really are.

And why?

Well such statements, the researchers suggest, our brains can swallow whole. We don't have to worry about chopping things up into smaller, more manageable pieces. We process such language more quickly. Such insight and meaning more fluently. And fluency, as we've just seen in politics, breeds confidence.

When I was a kid I remember the boxer Muhammad Ali prognosticating before a fight about which round he would win in. Funny thing was, he often did it in rhyme:

He hits like a flea so I'll take him in three.

He wants to go to heaven so I'll drop him in seven.

He thinks he's great so I'll get him in eight.

Was Ali, unconsciously or otherwise, tapping into a secret law of persuasion here? Did his ability to couple make the minds of his opponents more supple? Did his penchant for rhyme make him better at judging time? Did putting it in verse make his punches feel worse? It's certainly possible. A lot of Ali's predictions actually came true.

A few years ago now, when I first began work on split-second persuasion, I did the rounds among airline check-in staff. Purely in the interests of research, of course, I happened to ask them about first class. More specifically, how to get into it.

While boiling things down to a single "upgrade algorithm" proved impossible (well, I'm hardly going to tell you, am I?) a number of the people I spoke to mentioned humor. In fact, one Aer Lingus employee I interviewed in Dublin recalled an instance so extraordinary, an occasion when a line just so "worked" on him — "Do you have a window seat . . . *in first class?*" — that he didn't even have to think about it.

"It wasn't just what he said," the Aer Lingus guy told me. "It was the way he said it. I'm telling you — this bloke could've dealt crack at a Jehovah's Witness convention. It was the way he looked at me, even. It was, like, come on — I won't tell if you won't. He was confident, but not in a cocky way like you get with some of them. He was obviously chancing his arm, but it was so simple. I just wasn't expecting it."

And that's just it. The illusion that many of us have when it comes to persuasion is that it has to be complicated. It doesn't. Like the catchiest tunes — the ones that go round and round inside our heads — the catchiest influence is simple. It's cheeky. It's fresh. And it's right there in your face. Read back over what that Aer Lingus guy said. There's a telltale signature there. Incongruity. Confidence. Empathy. And, if you count the sneaky insertion of reciprocity — "I won't . . . if you won't" (more on that later) — perceived self-interest.

And all packed into nine simple words.

PERCEIVED SELF-INTEREST

PAIR OF BANKERS

The rock band Oasis were playing a gig in Manchester when technical problems forced them to leave the stage. When they came back, lead singer Liam
Gallagher announced to the 70,000-strong crowd, "Really sorry about that.
This is a free gig now. Everyone will get a refund." Next day, 20,000 fans
took him up on the offer — at a cost to the band of over £1 million. What
to do?

True to their word, Oasis coughed up. But there was a twist. Checks
were sent out personally signed by both Liam *and* brother Noel — and bearing a unique "Bank of Burnage" logo (Burnage being the area of Manchester
where the band started out).

A band spokeswoman said: "People can obviously cash them in. But
they are quite distinctive so a few people may decide to keep them."

A couple have shown up on eBay.

If you want the secret of persuasion in just a few simple words, it's easy.
Appeal to the other party's self-interest. Or, more specifically, to his *perceived* self-interest — what he *thinks* is to his advantage. It's also one of the
golden rules of management. Want to influence your boss? Then find out
what *her* boss wants. Quick recap: What's the best way of riding a horse?
Right. In the direction in which it's going. Hang around a schoolyard for
any length of time (or maybe, don't) and you'll soon see what I mean. Kids
get what they want from each other by doing one of two things. They either
trade (If you let me have a go on your PlayStation, I'll give you some of my
chocolate bar), or they make threats (If you *don't* let me have a go on your
PlayStation, I'll tell Mrs. Jenkins that you *stole* my chocolate bar). It's the law
of the jungle.

The smart ones even turn the tables on us grownups. I was at a New
Year's Eve party once when my friend was about to put her nine-year-old to
bed.

"But, Mum," he pleaded. "It's only eight-thirty. Let me stay up."

Mum was adamant.

"You know what you're like when you have a late night," she said. "You're
tired for days afterward."

"Well," replied her son, without missing a beat, "do you want me to be
running around at seven in the morning when you're having a lie-in?"

Nice.

Diplomacy, someone once said, is the art of letting other people have your way. And making sure they feel good about it.

PRESSING ENGAGEMENT

The Gallagher brothers aren't usually noted for their diplomatic skills. But over the refund incident they surpassed themselves. Chances are those checks will become collectors' items a few years down the line — and worth a heck of a lot more than they are right now. Cashed *or* flogged on eBay. Yet at the same time, no one can accuse the boys of not honoring their promise. Smart move.

What Oasis did here isn't rocket science. It's biology. By issuing those checks — the ultimate in "limited editions" — they were in fact tapping into an ancient law of influence called *scarcity*. Scarcity is one of six evolutionary persuasion principles outlined by our old friend Bob Cialdini, professor of psychology at Arizona State University — and refers to the observation that the less there is of something, the more we want it. The other principles, which we've already come across in various guises, include *reciprocity* (feeling obligated to return favors); *commitment and consistency* (like the Gallaghers, we aim to be true to our word); *authority* (we defer to those in power); *liking* (we say yes to those we like); and *social proof* (we check out what others are doing if we're not too sure ourselves).

Because of their evolutionary underpinnings, and of their role in primeval survival, each of these principles operates directly on the level of self-interest. Take social proof, for instance. A recent study from the University of Aberdeen reveals that if a man goes into a bar, his attractiveness level rises 15 percent if accompanied by a smiling female (take six and you can't fail). And exactly the same copycat reflex is found in animals. Female grouse and guppies — everything else being equal — choose mates they've previously seen copulating with a dummy over those they *haven't* (though I'm obviously not suggesting anyone goes *that* far). Why? Because under conditions of uncertainty or limited information, the principle of social proof acts as a powerful self-interest heuristic. If other females are attracted, then what's not to like?

The persuasive properties of self-interest are often difficult to convey on paper. None of us really like to think of ourselves in purely selfish terms — it's in our interests *not* to. But let's have a go with this.

Imagine that you, along with twenty-nine others, volunteer for a handsomely, if rather bizarrely, paid psychology experiment I'm running. When you turn up for the experiment, I show each of you into a separate cubicle which contains a buzzer conspicuously located on a central panel. I tell you, prior to your entering your cubicle, that you'll remain inside for a period of ten minutes, but that you're free to press the buzzer anytime you want. The first person to press the buzzer, however, will signify the end of the experiment. Oh, and one other thing. You can't communicate with anyone else in the cubicles.

Now I said that the terms weren't exactly straightforward — and here's where it gets interesting. If, at the conclusion of the full ten minutes, it transpires that no one has pressed the buzzer, then everyone — that's you, and your twenty-nine buddies — will each receive a free twenty-one-day vacation anywhere in the world of your choosing. However if, on the other hand, at any point during that ten-minute period someone does press the buzzer, then whoever it is will receive a free six-day vacation, and you and everyone else will get nothing.

The clock's ticking.

What do you do?

On first encountering the Wolf's Dilemma* — which, in case you were wondering, is what this is — most people don't need to be asked twice. It's plain as day what you should do. Hold out for the full ten minutes. If everyone sticks together, then everyone will be slapping on sunscreen for twenty-one days, right? Well, sort of. That's the big question, isn't it? *Will* everyone stick together? Maybe. Maybe not. What are the chances that there'll be someone in the group who, out of pure self-interest or random stupidity, "accidentally" presses the buzzer? Can you afford to take that risk?

The more you think about it, the more it becomes apparent that the rational option is, in actual fact, to press the buzzer yourself. If the chances are that at least one of the twenty-nine others is going to blow it for the rest — and jet off into the sunset for six days — then is there really any good reason why it shouldn't, in fact, be *you*? In fact, the best move of all is not even to *think* about pressing it. Just *press it*. As soon as you enter the cubicle.

---

*The Wolf's Dilemma was conceived by the American games theorist Douglas Hofstadter, *Metamagical Themas: Questing for the Essence of Mind and Pattern* (New York: Basic Books, 1985).

If, after all, *you've* managed to figure out that it's in your own best interests to hit the button, what's to say that one of the others won't have, too? And what's to say they're not going to do it *now*?

TEST OF TIME

It was the seventeenth-century British philosopher Thomas Hobbes who coined the phrase "a war of all against all" to describe what life would be like without government. (Though with the credit crunch, Afghanistan, and the MPs' expenses scandal one wonders if it's not worth a try.)

But I prefer the words of former Australian prime minister Gough Whitlam: "The punters know that the horse named Morality rarely gets past the post, whereas the nag named Self-interest always runs a good race."

Whitlam, in fact, might well have been speaking literally — if the results of a study at Princeton back in the 1970s are anything to go by. Psychologists John Darley and Daniel Batson divided students at Princeton Theological Seminary into two groups. The first group was told that they were to videotape a speech about the kinds of jobs they thought seminary students might be best suited to after graduation, while the second group was told to talk about the Good Samaritan. Both groups were then given several minutes' preparatory time in which to make some notes, after which the researcher informed them that the studio in which the recordings were to take place was situated in an adjoining building accessed by an interconnecting alleyway.

Now here comes the interesting bit. Before they set off, not all the students were given the same brief by the researcher. In fact, they were further subdivided into three additional groups and told completely different things.

The first group was told: "It'll be a few minutes before they're ready for you, but you may as well head on over. If you have to wait over there it shouldn't be for long."

The second group was told: "The assistant is ready for you, so please go right over."

And the third group was told: "Oh, you're late. They were expecting you a few minutes ago. We'd better get moving. The assistant should be waiting for you so you'd better hurry."

And then off they went.

The journey, however, had a surprise in store. Slumped in the doorway of the interconnecting alleyway was an associate. His head was down. His eyes were closed. And he wasn't moving. As the students passed by, he coughed — twice — and groaned.

The big question was this: How much help would each of the students give him?

To help them decide, the researchers agreed beforehand on a points system. The students would be awarded nothing if they failed (or appeared to fail) to even notice the "casualty." They'd get one point if they recognized his need for help but didn't stop. Two points if they didn't stop, but reported the matter to the videotape assistant who was waiting for them in the next building. And a maximum of five points if they stayed with the associate and accompanied him to a place where his condition could be assessed. The results of the study are shown in Figure 6.1.

Figure 6.1 — Can't stop — I'm late for my sermon.

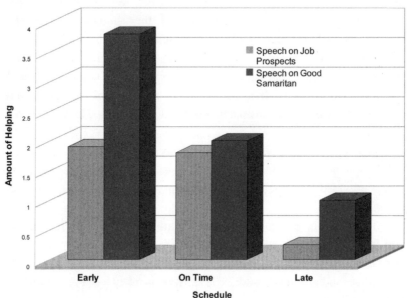

Sure enough, even among students at an elite theological seminary such as Princeton's, and even among students at an elite theological seminary such as Princeton's *who had just been making notes on the parable of the Good Samaritan,* self-interest told. And told big time. As the graph shows only too clearly, those students who were running late barely even noticed the man slumped in the doorway.

They were too busy trying to be good.

Early last year, the radio presenter Terry Wogan came under fire from various sections of the media after news leaked out of an £800,000 salary. In spite of the fact that his weekday slot drew in upwards of eight million listeners and had, over the years, afforded him iconic (some would say *ironic*) status, there were rumblings from various quarters that he was grossly overpaid. Add to that the prevailing financial climate, and an unsavory scandal involving two of his broadcasting colleagues, and the mud might well have stuck. But Wogan saw it differently.

"That," he quipped, with trademark avuncular aplomb, "works out at ten pence per listener. I think I'm cheap at the price."

End of story. As soon as self-interest was deflected, and the numbers rejigged to benefit his *detractors,* that, pretty much, was that.

INCONGRUITY

BRAIN SCAM

You don't find many conjuring tricks in science books. But here's one for you. In Figure 6.2a are six playing cards (see overleaf, top). Select one of them by touching it, stare at it for five seconds to build up a mental picture, and then briefly turn the book over and visualize it in your mind's eye. Don't tell me what it is, just keep it in your head.

Done that? OK, good. What I'm going to do now is pick the cards up and shuffle them, and then lay them back on the page — only this time face down and removing one in the process. You won't notice me doing this.

Ready? Here goes . . . (See Figure 6.2b, overleaf.)

Excellent! So far so good.

OK, what you should now have in front of you is five cards. Face down. The sixth card I have with me here. Bear with me for a moment while I just check to see which it is — I removed it without looking. OK, got it.

Want me to show you? To find out which card I'm now holding in my hand, go to p. 194 at the end of this chapter to "turn the cards over" for yourself. Do it now and come straight back.

So, which card was it? Was it, by any chance, the one you selected?

I guess that's magic, folks.

Figure 6.2a — Pick a card.

Figure 6.2b — Original display minus one.

PICKPOCKETING THE BRAIN

"The secret of conversion," the Greek philosopher Plato once wrote, "lies not in implanting eyes because the eyes exist already. Rather," he added, "it lies in giving the eyes a right direction which they have not."

He's spot on. Magicians, of course, have known this for centuries. As have pickpockets. "A big move covers a small move" is one of the most tried and tested maxims in the business — and refers to the fact that if two movements occur simultaneously, observers will attend to the larger, or more salient, of the two.

Take the "mind reading" trick above. You've probably gathered by now that it had nothing to do with mind reading whatsoever — and everything to do with mind *stealing*. With what in magic theory is known as *passive misdirection,* and in cognitive psychology as *exogenous attentional capture.* By getting you to focus exclusively on the one card — the card of your choice — chances are you hardly even noticed the other five. You were aware that they were there — you could *see* them — but you simply didn't *attend* to them.

Big mistake.

If you only have eyes for your own card, and have no idea about the identity of any of the others, all I need do is remove one at random and change the other five and it will appear as if the only card missing is your own. That one card — the card you chose — acts as a visual target. As a kind of "neural bouncer" — manhandling attention out through a side door of consciousness. And into a taxi home.

When it comes to persuasion we could learn a thing or two from magicians and pickpockets. Take Ron Cooper, for example, in the howling wind and rain. Irrespective of how hot they may be, who in their right mind would start taking their clothes off in *those* conditions? Would you? Cooper, of course, has a reason for what he's doing. The T-shirt. But the guy on the ledge doesn't know that. Instead, he must play along — as with each loosened button, the standoff gets more bizarre.

Then comes the grand finale: PISS OFF — I'VE GOT ENOUGH FRIENDS!

More incongruity. More psychological Semtex. Situations like this generally call for tact. For the shoulder-to-cry-on approach. Everyone knows that. The guy on the ledge knows that. And Cooper knows he knows it. But not this time. It's a risky play, but humor, Cooper has calculated, is a strong hand. Stronger (or so he hopes) than standing on the roof of a multistory car park in the rain.

Incongruity works in persuasion for precisely the reason that it *doesn't* work in magic. Because it's out of the ordinary. And yet, at the same time, for precisely the *same* reason. Big covers small.

STROOP TEASE

The power of incongruity to stop the brain in its tracks, to sneak up on it from behind and ram the barrel of surprise into its back, is nothing new. It's as old as the hills, in fact. "Make a noise in the east and attack from the west," the ancient Zen masters used to say — a doctrine still integral to many forms of martial arts today. In karate, for example, the concept of *teishin,* a "stopping mind," refers to a mind temporarily, and dangerously, dislocated from its primary focus. While in the courtroom — home, since the sophists of ancient Greece, to exponents of *linguistic* jujitsu — victory is also, in part, predicated on surprise.

The mercurial British barrister Frederick Smith once defended a bus driver against claims that his negligence had caused injury to a passenger's arm. Rather than employ an aggressive line of questioning with the plain-tiff, Smith, contrary to expectation, adopted a more conciliatory tone.

"Will you please show the court," he asked the passenger, "how high you're able to lift your arm now — *after* the incident in question?"

The respondent, doubled up with pain, raised his arm to shoulder level.

"Thank you," said Smith. "And now," he continued, "would you be so kind as to show the court how high you could lift it *before* the accident?"

The plaintiff's arm shot clean above his head.

The properties of distraction, which make incongruity such a force to be reckoned with in persuasion, may be appreciated in greater detail from the following exercise.

Take a look at the series of squares in Figure 6.3a (overleaf). In each of these squares, a word appears in a different location. Proceeding from top left to bottom right and working methodically along each row, your task is to say out loud the position each word appears in (left, right, up, down). State the position as quickly as you can. Don't read the words — just state the position in which each word appears.

With me? Let's go . . .

Figure 6.3a — Which position do the words appear in: up, down, left, or right?

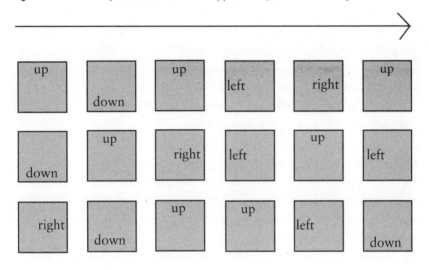

How was it? Pretty easy? OK, that's good.

Now what I want you to do is to repeat the task for the list of words shown in Figure 6.3b. Again—just state the position that each word appears in. DO NOT—I REPEAT, DO **NOT**—READ THEM!!

OK?

Off you go . . .

Figure 6.3b — Repeat as for Figure 6.3a. Which position do the words appear in: up, down, left, or right?

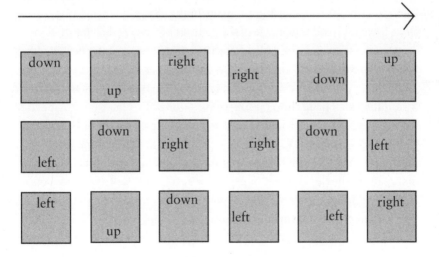

How did you get on this time? Different story? Thought so. Most people, in fact, find this second list considerably harder. But why? Well, the reason is really quite simple. On the second list the *conscious* instruction to *state the position* of the words crosses swords with the *unconscious* expectation of simply *reading* them — a juddering of the gears rendered particularly bumpy by the hideous incongruity between the words and their positions. Suddenly, in other words, expectation and reality are no longer an item. And performance on the task deteriorates.

This paradigm — a variant of something called the *Stroop Task* — is an old favorite of cognitive psychologists, especially those interested in the processes and mechanisms of attention. And with good reason. The *interference* or *disruption effect* generated by the two competing urges — one's natural inclination to read the words versus the diabolical directive to *override what comes naturally* and state their position — isn't just confined to language. It happens, in fact, pretty much all the time — whenever, for instance, we find ourselves in unfamiliar surroundings or are surprised by the unexpected.

Barbara Davis and Eric Knowles of the University of Arkansas have demonstrated how this works in a pair of studies involving door-to-door salesmen and street vendors. Davis and Knowles uncovered something remarkable about the way we spend our money: target clientele were twice as likely to purchase Christmas cards from a door-to-door salesman if he unexpectedly quoted the price in cents rather than dollars — and patrons of an outdoor market bought extra cupcakes from a stallholder if he referred to them not by their standard appellation but instead, somewhat unusually, as "half-cakes." But there was a catch. In both instances the sting only worked if a tagline was inserted immediately after the anomaly. In the case of the Christmas cards, "It's a bargain!" And in the case of the cupcakes, "They're delicious!"

Of course, what's going on here isn't particularly sophisticated. It's a low-level psychological con trick. The incongruous "first impression" — set of thirty-six Christmas cards for 2,844 cents — hustles the brain into skimping on the small print. Into "seeing cents" — or not, as the case may be. Before it has time to reconsider, confidence and empathy weigh in: "It's a bargain!" ICE (incongruity, confidence, and empathy) functions, as a unit, much like a SWAT team. The role of incongruity — 2,844 cents — is that of first man in: the deployment of explosive entry, and the creation of confusion. It induces in the recipient a momentary, split-second trance during which the confident, empathic nanohypnotic suggestion — "They're delicious!" — may be introduced covertly while persuasion has zero gravity and cognitive resistance is frozen. Hit the brain when its back is turned, and you can, quite literally, pretty much "name your price."

GOLDEN GAFFES

The neurology of incongruity—what happens inside the brain after the doors have been forced and the windows blown in—is actually pretty well documented. Single-cell recordings in monkeys have revealed that the amygdala is more sensitive to unexpected presentations of stimuli (both positive and negative) than expected, while in humans, intracranial EEG recordings have demonstrated increased activation in both the amygdala and the temporoparietal junction (a structure involved in novelty detection) on exposure to rare, especially bizarre, events. Indeed, as we saw in Chapter 2, incongruity, in the form of sudden and unexpected shifts in pitch, is what makes neonatal crying so emotive. And gives music and humor their edge.

But there is, as hinted at earlier by the cupcakes study, a secondary function of incongruity: one separate from, but not unrelated to, its facility for explosive entry. Its capacity to "reframe."

Take, for instance, the two advertisements shown in Figure 6.4—both very different, yet both, in their own way, very powerful. Adverts such as these are emblematic of what we might call guerrilla influence. They am-

Figure 6.4 — Veils of the unexpected: the power of disconfirmation in adversity.

bush expectation and take our emotions hostage. They force us to ask questions. To reevaluate.

Usually when people help us they have a smile on their face. Not cuts and bruises. What's going on?

Usually when a car hire firm takes out copy space it's to big itself up, not put itself down. What's the deal?

Drew Westen, professor of political psychology at Emory University, has the answer.

"If you want to win hearts and minds," he says, "start with the heart."

Westen and his coworkers have carried out a series of studies looking at the effects of just such emotional investment not in advertising, but in politics. What would happen, Westen wondered, if you took a group of diehard Republicans and a group of diehard Democrats and presented them with pairs of statements — pairs of *contradictory* statements — issued by their party leaders. In the case of the Republicans, George W. Bush; and the Democrats, John Kerry. Would it bother them? If so, *which ones* would bother them?

To find out, Westen did exactly that. In the run-up to the 2004 presidential election, he collated — from both sides of the party political divide — a selection of incongruous arguments (twelve in total, six vs. six — and we're talking real howlers here, not just minor inconsistencies) and showed them to both Republican and Democrat supporters alike, on a series of slides, while they reclined in an fMRI machine.

What he discovered was mindboggling. To neutral observers, the contradictions were obvious. As they were to the Republicans and Democrats — so long as they originated on the other side of the aisle. But could Republicans discern the inconsistencies in their own candidate's arguments? Likewise the Democrats? Not a chance. On a 1 to 4 scale (where 1 = not contradictory at all, and 4 = strongly contradictory), the average ratings for partisan statements hovered around 2. But for opposition statements — you guessed it — they were close to 4. Republicans, in other words, could only see the glitches in Kerry's rhetoric. And the Democrats, vice versa, in Bush's.

More importantly, however, was what was going on inside participants' heads while they were exposed to such dissonance. To begin with, as Westen and his team predicted, in the early stages of exposure (slides 1 through 3) incongruous ideology generated a burst of negative emotion in the brain (specifically, in the anterior cingulate, medial prefrontal cortex, posterior cingulate, precuneus, and ventromedial prefrontal cortex).

But as the experiment wore on (slides 4 through 6) something very interesting started to happen. The neural circuits involved in emotional regulation (the lateral inferior frontal cortex, inferior orbitofrontal cortex, insula, and parahippocampal gyrus) began to kick in. And kick in big time. And then, not only did the initial negative emotion start to dissipate, areas of the brain associated with *positive* emotion (the reward circuits of the ventral striatum) also showed up for the party. Participants didn't just start to feel *better*. They started to feel *good*.

No sooner, it seemed, had their brains recovered from the initial shock of the incongruity than they emotioned their way to reason. Somehow, they managed to reconcile the two conflicting statements. And rewarded themselves for doing so.

SHOCK AND/OR

The results of Drew Westen's study shed quite a bit of light on what goes on in persuasion. Not just in politics but in life in general. For starters, facts aren't always that important. As influencers they're overrated. When the chips are down and the going gets tough, the brain, it would seem, spends much of its time cowering behind the heart. Just to reiterate: all the action in Westen's study was in the brain's *emotional* zip codes. The *cognitive* neighborhoods were quiet.

In oratory, too, incongruity has a transformative quality. Contrasts such as those used by John F. Kennedy ("Ask not what your country can do for you, ask what you can do for your country") and Margaret Thatcher ("You turn if you want to, the lady's not for turning") make their points **in bold** because of the immediate juxtaposition of positive and negative. In fact, research has shown that of the total applause generated by a well-crafted speech, a third, on average, is triggered by this kind of symmetry.

Picture the scene as I jump on the New York subway one morning. Two beggars are sitting facing each other on opposite sides of the sidewalk. One is dressed in rags and is dejectedly holding up a sign which reads: *Hungry and Homeless — Please Help*. The other is dressed in an immaculate pinstripe suit and — grinning smugly — is brandishing a placard saying: *Filthy rich and want more!*

The reaction of passersby is intriguing. It's a mixture of disdain, sympathy, and amusement. As a marketing strategy, the guy in the pinstripe is a disaster. His bowl is pretty much as empty as when he started. Yet the "real" beggar — the guy dressed in rags — is raking it in.

I'm suspicious. There's definitely more to this little racket than meets the eye. So later, as they're packing up, I approach them. What's the story? I ask. Turns out I'm right. It *is* a racket.

In fact *both of them* are on the streets.

But they've discovered, through the psychology of what I think in the literature is called "balls," that by working together they can more than quadruple their day's takings.

"It gives people a choice," they tell me. "Rich guy or poor guy. Normally what happens — when you're on your own — is that people just walk straight by. They hardly even give you a second look. But the guy in the suit not only gets their attention, it gets them thinking. Why would I give anything to *him*, the smug bastard, when I can give it to the other guy? Regulars know it's a scam, but it still works. We take it in turn who wears it."

Big covers small.

CONFIDENCE

KEEPING WATCH

The story that I'm about to tell next doesn't exactly show my Uncle Fred in the best of lights. But it's such a great example of the transformative power of confidence that it's difficult to leave it out.

Fred Dutton served in the Parachute Regiment during the Second World War. He wasn't the biggest man in the world — around five feet six inches, and 9 stone wringing wet — but he was a tough little bugger and had the heart of a lion. One Christmas, somewhere deep in the Ardennes, he and three of his mates stumbled across a German position. Taken by surprise, the Germans decided to leg it — all except for the radio operator who couldn't get his backpack off in time. Fred was in charge and stepped forward.

"Stand up!" he yelled at the signaler.

The German did as he was told. He was six feet five inches, and built, as Fred was fond of saying, like a brick shithouse. For a few seconds the two men just stood there looking at each other. Quite literally, I guess, sizing each other up (or down, as the case may be). It must have been quite a scene: Fred at five feet six, Jurgen at six feet five. Eventually, Fred's eyes alighted upon the German's watch. It was shiny. And gold. And looked expensive. There was no doubt whatsoever in Fred's mind that his mates would back him up if things turned nasty. So what the hell? he thought. He may as well go for it.

"Your watch!" he barked at Jurgen. "Mine!"

As an aid to translation he pointed at the watch. Then he pointed at himself. The German looked at him as if he were mad.

"Your watch!" Fred repeated. "Hand it over!"

Still the German just stood there, eyeballing him ever more suspiciously as their bizarre standoff unfolded.

Finally, Fred had had enough. Walking right up to within a few inches of the German — horizontally, if not vertically — he gesticulated furiously at his wrist.

"YOUR WATCH!" he hissed. "GIVE IT TO ME!"

At the third time of asking the German, somewhat hesitantly, undid his watchstrap and handed the article over.

Fred grabbed it, shoved it into the pocket of his greatcoat, and with a big fat grin on his face turned round gleefully to his mates.

Who'd absconded, he later found out, as soon as they'd seen the size of the bastard.

OMNIA DICTA FORTIORA SI DICTA LATINA*

Reality, wrote the American essayist Robert Anton Wilson, is what you can get away with. And I'll tell you something for nothing: Uncle Fred would almost certainly have agreed with him. There was only one thing that convinced that German radio operator to hand over his watch — and it wasn't the festive spirit, that's for sure. Raw confidence was what it was. No doubt about it. Well, raw *misplaced* confidence actually.

Someone else who knows a thing or two about confidence is Greg Morant. It's a steamy summer's evening in New Orleans as we sit in the bar of his five-star hotel sipping champagne. "Persuasion," Morant tells me, in

---

*Everything sounds more impressive when said in Latin.

crisp white shirt, pale blue jeans, and with a gold Oyster Perpetual twin-kling on his wrist, "is 99 percent confidence and 1 percent coincidence!" And Morant should know better than anyone. Now in his midforties, he's been hustling for thirty years. And there's not a state in the Union where he hasn't quit while ahead.

"If you don't trust someone, if you don't have confidence that things will turn out as he says," Morant continues, "then what's the point in listening to him? Now that's no good to someone in my profession. Our bonds are our words! Did you hear the one about the confidence man who wasn't confi-dent? I mean, it's crazy."

He's right, of course. There's nothing like confidence to inspire confi-dence. Take TV, for example. If you've ever wondered why experts inter-viewed on television invariably appear against a backdrop of books — now you know. The accoutrements of knowledge lend their pronouncements that extra degree of oomph.

Or take Stanley Milgram's electric shocks experiment at Yale in the 1960s. A staggering 65 percent of those who took part in the study twisted that dial right the way round to maximum when instructed to do so by a be-nign-looking professor in a white coat. But when the professor shuffled off and a lab technician took over — in jeans, T-shirt, and sneakers — the "in-terrogators" weren't so keen. In a postscript to the original study, in which the stamp of authority and the cues of scientific rectitude were "dumbed down" (unlike the original study which was conducted in the hallowed en-virons of Yale University's "old" campus, the follow-up took place in an of-fice block downtown), only 25 percent of participants went the whole way. Still pretty shocking, but hey — not *as* shocking.

When confidence goes out of the window, everything else does, too.

A picture paints a thousand words, or so the saying goes. But is it pos-sible for a picture to say too much? On the surface, this seems like an odd question. But in the courtroom, there's evidence to suggest that the intro-duction of fMRI scans to the proceedings might actually be playing with fire. A recent study by David McCabe of Colorado State University and Alan Castel of the University of California in Los Angeles suggests that whatever benefits brain-imaging pictures bestow upon the legal process may in fact be outstripped by their in-built propensity to dazzle.

McCabe and Castel presented volunteers with a series of fictitious neu-roscience articles that included some dodgy reasoning (e.g., "watching TV helps with math ability because both activate the temporal lobe"). Whereas some participants received just the bad arguments, others received the arguments plus either brain images or bar graphs. Guess who thought the articles made more sense? You got it. Those who got the brain images.

Statistics, used well, convey the same psychological swagger. At the beginning of the O. J. Simpson murder trial in 1995, the odds of an acquittal seemed pretty slim. But a brilliant defense lawyer by the name of Alan Dershowitz had other ideas. Around four million American women were beaten up by their partners each year, he confidently submitted to the court. But out of those four million only 1,432 (in 1992) had actually been *killed* by their abusers. Given these figures, Dershowitz argued, it followed that the odds on his client being guilty were in fact around 1 in 2,500.

The jury was impressed by Dershowitz's fiendish arithmetic. And Simpson, after a trial lasting 251 days, walked out of court a free man.

But the math turned out to be *wrong*.

And the data, unbeknown to the prosecution, concealed a radically different possibility. Since Nicole Brown Simpson was *already* dead, Dershowitz's odds were facing in the wrong direction.

Of those 1,432 women who'd been murdered, *90 percent* had died at the hands of their partners.

RING OF CONFIDENCE

Psychologist Paul Zarnoth and his coworkers at the University of Illinois have examined the effects of confidence on cognitive function. More specifically, how an aura of self-assurance can envelop those around us in a vicarious cloud of truth. Zarnoth presented volunteers with various kinds of problems (e.g., math, analogy, and forecasting tasks) and asked them, after each one, to indicate how confident they were that they'd got the answer right. Volunteers responded first as individuals, and then again in small groups. In neither case did they receive any feedback about their performance.

Zarnoth's results were extraordinary. Group responses, he discovered, appeared to follow a pattern. They tended to mirror the *individual* responses of the most confident group members — even when they happened to be wrong. In other words, concluded Zarnoth, those individuals who were perceived as being the most *confident* were also the ones who were perceived as being the most *competent*— the most likely to have answered correctly.

And it doesn't take much to pick up on confidence. Surprisingly little, in fact. In politics, studies have shown that one of the strongest predictors of candidate popularity is approach behavior; when, during a question-and-answer session for example, the candidate moves *toward* an audience (exuding a subradar confidence and openness) as opposed to remaining stationary (signifying defensiveness).*

Psychologists Nalini Ambady and Robert Rosenthal have taken things one stage further, and conducted research into something they call "thin slicing." In one study, raters viewed thirty-second video clips of college lecturers at the beginning of an academic term and evaluated them on a number of personality variables. Would these minimal evaluations (or "thin slices"), Ambady and Rosenthal wanted to know, predict how the lecturers did at the end of term — some three months down the road? Not, that is, in the eyes of the raters, but in those of the lecturers' students.

Remarkably, so it proved. Those lecturers initially perceived as confident, active, optimistic, likeable, and enthusiastic — after just thirty seconds, remember — fared far better on student evaluation forms later in the year.

Oh, and I forgot to mention . . . even more remarkable was the fact that raters gave their initial ratings with the sound turned off. The videotapes had no audio. All participants had to go on was the evidence of their eyes.

Confidence, like physical attractiveness, gives off a halo effect. It acts unilaterally as a stand-alone marker of influence.

If it didn't, then good old Uncle Fred would have been long gone.

EMPATHY

SMOKE WITHOUT FIRE

It's a Friday night and the London Underground is busy. For five minutes or so a Piccadilly Line train has been backed up in a tunnel between Leices-

---

*In medicine it works the other way. The extent to which nurses are able to inhibit facial expression and conceal their inner feelings is correlated with higher ratings from their superiors (perhaps not surprising given the occasional necessity to hide from patients the true severity of their condition).

ter Square and Covent Garden because of a signaling failure. The carriages are full and people are getting restless. Even more so now that the driver has just announced a further five-minute delay.

A guy in a tracksuit takes out a cigarette and lights it — a real no-no.

Ever since the King's Cross inferno of 1987 which claimed the lives of thirty-one people, and which, a subsequent inquiry revealed, was started by a discarded match, there's been a blanket smoking ban on the Underground. But despite the fact that there are No Smoking signs all over the place, the guy lights up anyway.

An uncomfortable silence descends upon the carriage. The looks on people's faces speak volumes. But no one — as so often proves the case — breathes a word. Then, out of the blue, a guy in a pinstripe suit breaks the curfew.

"Excuse me," he says, leaning forward with a cigarette, "you couldn't, by any chance, give me a light, could you?"

This, it transpires, is the final straw. Immediately, another passenger intervenes.

"You DO KNOW that you can't smoke in here?" he snaps.

The guy in the suit suddenly "notices" the No Smoking signs.

"Sorry," he says. "Didn't realize."

Then he turns to the guy in the tracksuit.

"Maybe," he says, "we'd better put them out."

We've all been in situations like this, haven't we? And often, unless you happen to be made of sheet metal, the right course of action isn't immediately obvious. Someone who's prepared to nonchalantly light up a cigarette in a prohibited zone is not, in all likelihood, going to "come quietly." He's going to put up a fight.

So what, in this case, does our fellow passenger do? Well, rather than choosing the usual route of overt confrontation, he goes, instead, for the complete opposite. In direct contrast to what the guy in the tracksuit is expecting (a challenge) he *joins* him instead ("Got a light?") — knowing full well, of course, that such collusion is bound to provoke a reaction from one of the other people in the carriage. Which it does. But by the time that reaction materializes, the game's already up. Crucially, there's now no longer just the one transgressor in the frame, but *two*. The picture has changed dramatically. Suddenly, in the blink of an eye, a makeshift "in-group" has been created and there's safety in numbers.

The optimal outcome — getting the guy in the tracksuit to put out his cigarette — may be framed in terms of a friendly request from a fellow "unwitting offender" rather than as a full-on challenge from on high.

And things return to normal.

FACING FACTS

If the ingredients of effective persuasion were to be arranged like poker hands in order of strength, then the way we *feel* about people would often rate higher than what they say or do. Take what just happened on the London Underground. The reason why the guy on the train extinguished his cigarette wasn't because he was *told* to extinguish it (although that obviously had something to do with it). It was, instead, because of the *way* he was told. And by whom. Such an ability to customize a message so as to maximize its impact, to serve it up "warm" to whoever it is meant for, calls for empathy — for a good working knowledge of the principles of emotional synchrony — and there are two main ways you can do it. First, you can reduce the psychological distance between yourself and the recipient: increase *similarity*. Or you can frame what you say in such a way as to make it more "personal": increase *salience*.

"Wanna get a kid to eat potatoes?" laughs master con man Greg Morant, as the slow Southern twilight glints off his Rolex and he orders us more champagne. "Then serve 'em up as fries."

Lisa DeBruine at the University of Aberdeen's Face Research Lab has conducted a fascinating investigation into the mechanics of similarity. More specifically, into its effects on trust.

What she did was as follows.

First, DeBruine devised a computer game for "two" players. In the game, each individual player was presented with a choice. The players could either:

1. divide up — between themselves and their partner — a *small sum of money personally*, or
2. trust their *partner* to distribute a *larger* amount.

The participants were assigned sixteen different partners whose faces were displayed on a monitor. But there was a catch. Unbeknown to the participants, all of the "partners" who appeared on the screen in front of them had had their faces altered using a face morphing technique (see Figure 6.5). None of them, in other words, were "real."

But that wasn't all. While half of these partners comprised a composite of two *strangers*, the other half was different. In this other half, it was the participant's *own* face that had been morphed with the stranger's.

What, DeBruine wondered, would be the pattern of participant choice? Would they, as predicted by the principles of kin selection, be more willing to cede control to their "playing partners" when that partner more closely resembled themselves? Or would facial similarity have little bearing on trust?

The results were pretty impressive. On average, DeBruine found that participants trusted players with faces resembling their own on more than two-thirds of the trials — compared with only half the time when the face on the screen was completely unfamiliar.

Seeing ourselves in others can sometimes come at a price.

Figure 6.5 — Face morphing. Participants' faces (left) were morphed with the faces of strangers (right) to generate the composite faces (center). The top and bottom panels provide examples of different gradations of morphing. The female morph (top) assimilates both shape and color information from the participant and unfamiliar face, whereas the male morph (bottom) assimilates only shape information from the unfamiliar face.

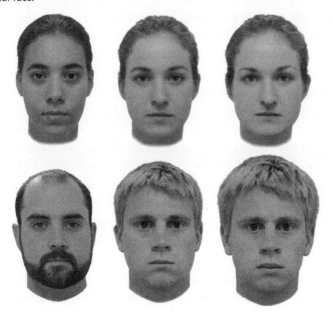

COMMON TOUCH

In the worlds of sales and marketing DeBruine's revelations wouldn't raise many eyebrows. In crucibles of influence such as these it's been known for quite some time that similarity is the name of the game. And you know what? It doesn't matter *where* the overlap starts. It's just enough that there *is* one.

An amusing study which demonstrates this divided a bunch of students into two groups: those who were told that Grigori Rasputin — the notorious "Mad Monk of Russia" — happened to share the same birthday as theirs, and those who were told that his birthday fell on a different day. Each group then read an account of Rasputin's dastardly deeds and was subsequently asked to rate how "good" or "bad" he was. Though Rasputin's heinous CV was identical for both groups, which of the participants do you think rated him more positively? Correct — those who "shared" his birthday.

The results of studies such as these have strong implications for the way we influence others. On January 20, 2009, in the eighteen minutes and twenty-eight seconds that comprised Barack Obama's inaugural address as president, those key, umbilical utterances "we" and "us" and "our" tumbled like sparks from his great oratorical anvil — appearing, between them, a total of 155 times in the speech. "We're all in this together" was the subtext. The main text, even. Yoked one to another by history — by the *Mayflower,* by Gettysburg, by 9/11 — Americans face the future side by side.

More circumspect was Obama's allusion to the "young preacher from Georgia" in his Democratic nomination speech. By not referring to Martin Luther King by name — a rhetorical technique known as *antonomasia* — he crafts, at a stroke, an effortless intimacy between speaker and audience: the flattering assumption that we're all "insiders" here, we all know who I mean. And look how the term humanizes King: there was a time, before he turned into stained glass, that he, too, was just an ordinary Joe like us. And how the mention of Georgia (a favorite trick of Obama) grounds and localizes the rhetoric: "Our campaign . . . began in the backyards of Des Moines and the living rooms of Concord and the front porches of Charleston."

Politicians and salespeople make a point of drumming up empathy, of staking their claim to oases of common ground, with good reason. Because it works. And the better the location for that common ground — the more meaningful or "prime" it happens to be — the more we want to buy into it.

Once, going in to buy some shoes in Lexington, Kentucky, I ran into a sales assistant who grew up just two streets away from where I did — five thousand miles away in a tiny corner of West London. What a coincidence *that* turned out to be. I felt I almost *had* to buy a pair of shoes off him. And did. Two pairs, in fact. Couple of days later both were in the bin.

And it doesn't just work commercially. Not long ago, on a flight to New York, I sat next to a young, cool-looking guy of around twenty-five who wasn't in the least bit concerned that he had no street address to put down on his immigration form (such as is now the requirement).

"No problem," he said. "Just watch."

I wasn't so sure. Standing behind him in the queue at JFK, I was all ears. Was he really going to pull it off? If so, how? There was the usual exchange of insults with the officer on the desk as she took his prints and photograph. But then, when she eventually got round to processing his form, he suddenly remarked on her name.

"Wow — Verronica with two r's! That's amazing! The only other person I've ever known spell their name like that was my mum. That's great!"

The officer beamed. She agreed it was quite a coincidence.

Know what? First time she, too, had come across another Verronica. She stamped his passport. Handed it back to him. And that, pretty much, was that.

Bit of distraction. Bit of empathy. And he was in.

SIXTH SENSEI

I have absolutely no doubt that there are mind-reading geniuses out there because I've met one of them. In the higher echelons of the martial arts there's a test. The test involves one man kneeling down — arms by his side, blindfolded — while another stands behind him with a samurai sword raised aloft. At a moment of his choosing, the man standing behind will bring down the sword onto the kneeling man's head causing severe injury, possibly death — unless, that is, the blow is somehow deflected and the swordsman subsequently disarmed.

Such a feat appears impossible. And yet, it isn't. What I have described is real: an ancient, impeccably choreographed test, carried out in remote *dojos* in Japan and the High Himalaya, that those approaching greatness — those sinewy sorcerers miles above the black belt — have to pass. These days, thankfully, the sword is made of plastic. But there was a time, many years ago, when it was the real thing.

An ancient *sensei,* way into his eighties, told me the secret.

"One must empty one's mind totally. One must focus purely on the now. When one enters a state like that, one is able to smell time. To feel its waves washing over one's senses. The tiniest ripple may be detected over great distances, and the signal intercepted. Often it appears that the two combatants move simultaneously. But this is not so. It is not difficult. With practice it can be mastered."

Equivalent empathic genius may also be attainable aurally, in the realm of linguistics. A prostitute of over twenty years' experience told me — strictly in the line of business — that she was able to gauge, within thirty seconds of talking to him on the phone, whether or not a client posed a risk. Whether, in other words, it was safe to invite him round.

"I can't explain it," she said. "It's just something you pick up. And in this business it's something you *need* to pick up — it could be the difference between life and death. When I first started out, I used to get knocked about quite a bit. But now it never happens. Soon as I hear a voice I start to build up a picture. I get a vibe. It's like a sixth sense. And its very rarely wrong."

Most of us are never going to be this good at hacking into other people's minds. Because most of us, thankfully, are never going to *have* to be this good. But here's the deal. To influence others, you don't need to be an expert mind reader. Sure, we each have our own *individual* frequencies on which the signal comes through at its sharpest. But there's also a network of shared wavelengths which all of us tune into.

The importance of nailing the right frequency, the appropriate psychological bandwidth on which to transmit our message, is demonstrated in a study conducted by Victor Ottati at Loyola University and his coworkers at the University of Memphis. The study, ostensibly at least, examined the benefits of graduate theses requirements — but was, in reality, all about figurative language. Ottati took a bunch of messages with built-in sports metaphors (e.g., "If college students want to play ball with the best, they shouldn't miss out on this opportunity") and compared them with a bunch of neutral messages (e.g., "If college students want to work with the best, they shouldn't miss out on this opportunity"). Which of these two message types, Ottati wanted to know, would spark the greater interest? And which would be considered by students as exerting the greater influence?

The results were unequivocal. Analysis revealed that the messages containing the sports metaphors were not only processed more carefully — they also, subsequent to evaluation, had the greater impact on attitudes.

But — and here's the key — only for students who were sports fans. For those who weren't interested in sport, the metaphor blew up in their faces: attenuating interest in the issue of thesis requirement and considerably reducing persuasion.

"The orator," observed Aristotle in the fourth century BCE, "persuades by means of his hearers, when they are roused to emotion by his speech; for the judgements we deliver are not the same as when we are influenced by joy or sorrow, love or hate."

Nowadays, of course, the data proves him right. Not least from fMRI.

Recall, for example, Drew Westen's "political divide" study from earlier. Westen showed us that if we happen, at the outset, to have a strong political allegiance to one particular party or another, then no amount of wrangling will get us to change our minds.

The brain, intoxicated by empathy, simply chokes on its own logic.

Recently, the CIA has discovered a secret weapon in the war on terror: Viagra. Many Afghan warlords have half a dozen wives, and maybe, someone realized, could do with a little assistance. One official recalls a sixty-year-old tribal leader welcoming him with open arms having received, on a previous visit, a little box of tablets.

"He came up to me beaming, saying, 'You are a great man.' Afterward, we could do whatever we wanted in his area."

Simplicity, perceived self-interest, incongruity, confidence, and empathy: if it can stiffen resolve against the Taliban, just think what it could do for you.

SPICE OF LIFE

"Why lie? I want beer!"

"You calling me fat?"

PISS OFF — I'VE GOT ENOUGH FRIENDS!

These three examples of what I call split-second persuasion each have something in common. All of them — yes, even the last one — elicit strong positive emotions. As instruments of influence this bodes well. Research has shown that one of the biggest predictors of altruistic behavior is current mood state — how you feel at the time. Feel good and the beggar gets lucky. Feel lousy and you scurry straight past.

Even when it comes to putdowns, the feel-good factor is key. When the legendary Australian fast bowler Glenn McGrath asked the Zimbabwean batsman Eddo Brandes how come he'd gotten so fat, Brandes, with characteristic decorum, replied rather eloquently that each time he screwed McGrath's girlfriend she obligingly gave him a biscuit.

Even the Australians laughed.

In fact, they gave him a round of applause.

Persuasion that makes you *like* it, makes you *do* it.

The feel-good factor inherent to split-second persuasion is enshrined within each of its modules. For some — confidence, empathy, perceived self-interest — this may perhaps seem rather obvious. But for simplicity and incongruity, the evidence is just as compelling. Research, for example, using facial electromyography (EMG) has demonstrated a direct correlation between the fluency with which a stimulus is processed and increased activation of the zygomaticus major — or "smiling" — muscle. Moreover, when a stimulus is processed unexpectedly fluently (think: dropping in on a neighbor at home versus bumping into him at the theater) the tremors of positive emotion — the feeling of familiarity — reverberate even deeper.

This, in split-second persuasion, is why humor is often effective. When someone endeavors to change something about us — and our minds are no exception — the process, more often than not, turns out to be less than pleasurable. But if, on the other hand, the process turns out to be smooth — and in some cases, even enjoyable — "Why lie? I want beer!" versus "Vietnam vet . . . six months to live" — then we're not only likely to see where that person is "coming from," we're likely to want to go visit.

Orthographically, the position of incongruity at the center of the SPICE model is also reflected dynamically. From calming someone down to raising someone's spirits, from closing a deal to trying to bum a quarter on the street, defiance of expectation, script reversal, antithesis — call it what you will — lies right at the very heart of split-second persuasion. Not only, as we've just noted, does it enhance the aesthetic prowess of simplicity, it also, as we saw earlier, knocks out the brain's surveillance mechanisms — enabling the rest of the SPICE model taskforce to secretly slip in subradar; to hijack resistance and hot-wire our neural pleasure centers.

The effects are irresistible. The result is a persuasion that doesn't just take the brain hostage. It actually makes us *not* want to pay the ransom.

Think *flipping* point rather than tipping point.

Mind-*jacking* instead of mind-hacking.

Persuasion as nature intended — before language dumbed it down.

During the Second World War, German bombers were an all-too-familiar sight in the nighttime skies over London, and parts of the city were completely razed to the ground. One area particularly badly hit was the East End.

One morning, after a not untypical night before, Whitechapel High Street lay in ruins. As, one might have expected, would the spirit of its residents. Not so. In the window of a grocer's store — in fact, in the only pane of glass still left in the building — the proprietor had posted the following:

IF YOU THINK *THIS* IS BAD, YOU SHOULD SEE OUR BERLIN BRANCH!

Indomitable. Inviolable. Irresistible.

The SPICE of life, it's called.

SUMMARY

In this chapter, we decoded the secret structure of persuasion. We sequenced the genome of the most powerful strain of influence on the planet, and uncovered a nucleus comprising five core factors. These factors (simplicity, perceived self-interest, incongruity, confidence, and empathy, or SPICE for short) are persuasion at 10, when it's usually at 6 or 7 — and when deployed in unison, dramatically increase the chances of us getting what we want.

In the next chapter we turn our attention away from theoretical models — and on to individuals. So far in this book we've met a number of people for whom persuasion is a profession: for some the kind that pays a monthly salary; for others the kind that nets illegal millions.

What is it about these latter individuals that sets them apart from the rest? That enables them to breach our brains' most sophisticated surveillance systems without us even noticing?

The answer may surprise you.

Prepare to meet . . . the psychopath.

Card Trick — Which is the missing card?
Could it be the one that you chose?

7

The Psychopath —
Natural Born Persuader

He couldn't have given a damn about anything. He took everything in his stride. And whenever anyone had a problem — like with his wife or girlfriend or something — he'd get to the heart of it in seconds. He exuded a kind of laser psychology. Like he could break into your brain and you didn't know he was there. If I hadn't seen him slit a man's throat and smile as the blood oozed between his fingers, I'd have said he was fucking Jesus.

— Special Forces sergeant on a former troop member

I can read your brain like a subway map. Shuffle it like a deck of cards.

— KEITH BARRETT

THE ONE

Secure Unit, Summer 1995

"What are you doing tonight?"

"Don't know. Going out probably. Pub. Club, maybe? Why?"

"What are you going to do there?"

"What do you mean, what am I going to do there? Usual stuff, I suppose. Meet up with some mates. Have a few beers . . ."

"Pull some birds?"

"Yeah, I guess. If I'm lucky."

"And what if you're not?"

"Not what?"

"Lucky."

"There's always next time."

He nods. Looks down. Looks up again. It's hot. This is a place where the windows don't open. Not because they won't, but because they can't. Don't try to outsmart him, the psychiatrist had said. You've got no chance. Your best bet is just to play it straight.

"Do you think of yourself as a lucky person, Kev?"

I'm confused.

"What do you mean?"

He smiles.

"Thought so."

I swallow.

"What?"

Silence. For about ten seconds.

"There's always one, isn't there, Kev? The one you think about as you're eating your hot dog on the way home. The one that got away. The one you 'never got round to' because you were just too fucking scared. Scared that if you *did* get round to her you'd end up doing exactly what you end up doing every other Friday night. Eating shit. Talking shit. Feeling shit."

I think about it. He's right. The bastard. Sort of. A sea of faces strobes across my brain as I stand in the middle of an empty dance floor somewhere. Anywhere. What am I doing there? Who am I with? The promise of emptiness yanks me back to the present. How long have I been gone? Five, ten seconds? I need to respond. And fast.

"So what would *you* do?" I say.

Pathetic.

"The business."

No hesitation.

"The business?" I repeat.

I'm on the ropes here.

"And what if she's not interested?"

"There's always later."

"Later? What do you mean?"

"I think you know what I mean."

Silence. Another ten seconds. I *do* know what he means and it's time to wrap things up. I rummage around in my briefcase and power down the laptop. A nurse looks in through the glass.

"Mike," I say, "it's time for me to check out. It's been good talking to you. I hope things go OK for you in here."

Mike gets up. Shakes my hand. Coils his arm gently around my shoulders.

"Look Kev, I can see that I've offended you and I really didn't mean to do that. I'm sorry. Enjoy yourself tonight. And when you see her — *her,* you'll know who she is — think of me."

He winks. I feel a pulse of affection and am filled with self-loathing.

I say: "I'm not offended, Mike. Really, I mean it. I've learned a lot. It's brought it home to me just how different we are. You and me. How differently we're wired. It's helped. It really has. And I guess the bottom line is this: that's why you're in here and I'm (I point at the window) out there."

I shrug, as if to say it's not my fault. As if, in a parallel universe, things could just as easily have turned out different.

Silence.

Suddenly, I'm aware that there's a chill in the room. It's physical. Palpable. I can feel it on my skin. Under my skin. All over me.

This is something I've read about in books. But have, up until this moment, never experienced.

I stand for five agonizing seconds in a stare forty below. Ever so slowly, as if some new kind of gravity has been seeping in unnoticed through the vents, I feel the arm vacate my shoulders.

"Don't let your brain piss you about, Kev. All those exams — sometimes they get in the way. There's only one difference between you and me. Honesty. Bottle. I want it, I go for it. You want it, you don't.

"You're scared, Kev. Scared. You're scared of everything. I can see it in your eyes. Scared of the consequences. Scared of getting caught. Scared of what they'll think. You're scared of what they'll do to you when they come knocking at your door. You're scared of *me.*

"I mean, look at you. You're right. You're out there, I'm in here. But who's free, Kev? I mean, *really* free? You or me? Think about that tonight. Where are the *real* bars, Kev? Out there" — he points at the window — "or in here?" (He reaches forward and, ever so lightly, touches my left temple.)

SUPERSANITY

Deep in the neurobiological cosmos, the brain of the psychopath is glimpsed in remote orbit, a moonless world of glacial desolation and eerie mathematical charm. No sooner is the word out than images of serial killers, rapists, suicide bombers, and gangsters come scything across our minds.

But what if I were to paint you a different picture?

What if I were to tell you that the psychopath who rapes your girlfriend might also, on another day, be the person most likely to rescue her from a burning building? Or that the psychopath of today who lurks with a machete in a dimly lit parking lot may well be the Special Forces hero of tomorrow — using that very same weapon in hand-to-hand combat in Afghanistan? Or that the supercool emotional assassin, the shadowy ambassador of charm whose mentholated morals and lightning sleight of mind robbed you blind of your entire life's savings, might also, if he put his mind to it, save you from going under?

Claims like this stretch credulity to the limit. And yet they're true. Unlike their box-office counterparts, not all psychopaths are violent. Far from it. Merciless and fearless, maybe. But violence is on a different neural freeway. One which sometimes has an interchange with psychopathy, but which just as often passes overhead.

Then, of course, there's charisma. The psychopath's famous "presence." Devastating. Dazzling. Disarming. These are the kinds of character references one often hears about such individuals. Not, as one might expect, from *themselves* but from their *victims*.

The irony is plain as day. Guys like these (and it usually *is* a guy)* appear, through some ghoulish trick of nature, to possess the very personality characteristics that many of us would die for. Indeed, that many who've fallen under the spell of a psychopath *have* died for.

They possess immense poise under pressure — their refrigerated hearts barely skipping a beat under conditions of even the gravest danger. They are charming, confident, ruthless, and remorseless. And out totally for themselves. No matter what.

They are also the kings of persuasion.

WILL THE *REAL* PSYCHOPATH PLEASE STEP FORWARD . . .

Let's get one thing straight right from the start. Being a psychopath doesn't make you a criminal. Not by default, at any rate. And it doesn't make you a serial killer either. In fact, many psychopaths aren't even in prison — they're

---

*The incidence of psychopathic disorder in the male population is estimated to be around 1–3 percent. In females, it's around 0.5–1 percent.

out there locking *others* up. This often comes as a surprise to many people, but it's true. A bit like the areas on a subway map, there are, in fact, inner and outer zones of the disorder, with only a small minority resident in the "inner city." Psychopathy is a spectrum along which each of us has our place. And just like any scale or dimension, it's going to have its A-listers.

The assumption that there exists a dichotomy between psychopath and nonpsychopath is descended from the lineage of clinical diagnosis — more often than not within a forensic setting — using standardized psychometric scales. The Psychopathy Checklist–Revised (PCL-R) is a specialized, and well-established, questionnaire originally developed for clinical use by the Canadian psychologist Robert Hare. It measures core psychopathic traits such as charm, persuasiveness, fearlessness, lack of empathy, and absence of conscience. On its 40-point scale, members of the general population will typically register around 4 or 5 — whereas a score of 30 is generally considered the entry level for psychopaths.

In clinical settings, as indicated by performance on the PCL-R, "home run" psychopaths — the Mikes of this world — knock the ball clean out of the stadium. There is, no question, a world of difference between these guys and the rest of us. But the problem with this is that we don't all live in rarefied clinical settings. And while Hannibal Lecter is serving up liver for breakfast, the traits that mark these "pure" psychopaths out from the rest of us happen, like personality traits in general, to be distributed evenly across the general population as a whole. Just as there's no official border between someone who plays the piano and a concert pianist, or between someone who plays tennis and, say, a Roger Federer or a Rafa Nadal, so the frontier between a "world-class" psychopath and someone who merely "psychopathizes" is similarly blurred.

Think about it. One individual, for example, may be extremely cool under pressure, and demonstrate a sublime lack of empathy (and we'll see a little bit later how *that* pans out on the trading floor), but at the same time act neither violently, antisocially, nor without conscience. High on two psychopathic characteristics, he or she might therefore be considered further along the "psychopathic spectrum" than someone scoring lower on those two traits, but wouldn't fall within the "danger zone" of someone scoring high on all of them.

Like the dials on a studio mixing deck, the "soundtrack" of personality is graded.

Psychologists Scott Lilienfeld and Brian Andrews have devised an alternative test to the PCL-R based precisely on such a soundtrack. Better suited to detecting the presence of psychopathic traits among the nonclinical population (in those not so much behind bars as propping them up), the Psychopathy Personality Inventory (PPI) provides a more sensitive measure of psychopathic traits: the existential emphasis being predominantly on psychopathy as a continuous predisposition rather than a stand-alone disorder.

This, of course, has profound implications for the way we approach the condition and raises some important questions.

Is psychopathy an all-or-nothing affair? Or more like a virus, where we can "test positive" in the lab yet not exhibit the full-blown array of symptoms? Are psychopaths qualitatively different from the rest of us — or just at the deep end of a rather murky gene pool?

And is it possible that far from posing a danger to individuals, or society in general, psychopaths might actually have something *special* to offer: that the right combination of psychopathic traits, sampled and mixed at carefully calibrated "volumes," might put us ahead of the game?

It's this latter observation that makes the psychopath such a mystery to scientists. When, as a postgrad, their subemotional profiles first appeared on my radar, it was the psychopath's psychological cat-burgling skills that intrigued me most of all. A core trait measured by all indices of psychopathy is the ability to persuade: the capacity to influence others.

Yet here's the deal. These measures, as we've seen, also gauge levels of *empathy*. Which is odd.

How, I wondered, could anyone lacking empathy be so brilliant at social influence? Psychopaths are recognized as being the best in the business at knowing what makes us tick. At getting under our skin. At getting inside our heads. Take Keith Barrett, for instance. Or Mike, whom you've just met. Mike had raped eight women, and killed two. He was, as the psychiatrist alluded to, a real-life Hannibal Lecter. A psychological black belt not to be messed with, which I discovered to my cost.

But in order to run their software — their programs of persuasion — Keith and Mike must first have acquired the hardware. And not just any old hardware — the hardware of empathy. Difficult to come by if you're a psychopath.

Suddenly, it got me thinking. If SPICE really *was* a universal model of influence, then what, precisely, were the psychopaths up to?

HIDDEN SHALLOWS

The advent of sophisticated brain-imaging techniques such as functional magnetic resonance imaging (fMRI) and magnetoencephalography (MEG) has sometimes been likened to the lunar landing. Finally, at our fingertips, we have the technology to launch ourselves not into *outer* but *inner* space. That will allow us to "land" on that mysterious gray planet that each of us knows so well—yet few of us, up until now, have ever properly explored: the world between our ears. But some worlds, quite clearly, are more hospitable than others. And some, just like their cosmological counterparts, seem far better suited to supporting life than others. Some are warm, bright, and quite easily habitable. Some seem polar, dark, remote—barely recognizable in the icy outer fringes of the neurobiological firmament.

One such world is the world of the clinical psychopath.

It's often difficult to appreciate the sheer magnitude of difference between the world of the "pure" psychopath—the psychopathic A-lister—and the nonpsychopath. Former U.S. Marine and nightclub bouncer David Bieber "calmly" gunned down a traffic cop with a single shot to the head as the terrified constable, bloodied and badly injured, pleaded for his life just inches away from the trigger. A police-car radio picked up the officer's final, desperate words: "Please, don't shoot me. No . . ." before Bieber opened fire.

Committing him for sentence, the trial judge told Bieber he had shown "no remorse or understanding of the brutality of his crime" and that he continued to maintain a "cool and detached" demeanor when attempting to explain the evidence against him.

Another psychopath, twenty-four-year-old Tara Haigh, handed a life sentence in 2008 for smothering her three-year-old son to death with a pillow, was on an internet dating site just hours after the murder. She posted a message on the site saying her son had died from a tumor behind the ear—then proceeded to fix up a rendezvous.

This, in case you didn't know, is the kind of people we're dealing with.

Examples such as these, so far beyond the pale of normal human experience as to defy comprehension, provide graphic depictions of the psychopath's lack of empathy.

Or do they?

Actually, research suggests that there may be a bit more to it. And that, far from being an open-and-shut case, the question of whether psychopaths lack empathy or not depends, in fact, on what kind of empathy we're talking about.

There are two types: "hot" and "cold."

Hot empathy involves *feeling*. It's the kind of empathy that we "feel" when we see others performing a task, and appropriates precisely the same "shared" somatosensory brain circuits — plus the amygdala (the emotion-processing area of the brain) — as those that become active when we perform that task ourselves.

Cold empathy, in contrast, involves *calculation*. It refers to the ability to gauge, cognitively and dispassionately, what another person may be thinking and involves completely separate pieces of neural circuitry: primarily, the anterior paracingulate cortex, the temporal pole, and the superior temporal sulcus.

There's a world of difference.

Hot empathy without cold empathy is the meter without the verse. Cold empathy without hot empathy is the verse without the meter: the total opposite. It's like having a highly detailed map without the firsthand experience of what the symbols on that map relate to. You can still read it, still get around. But it doesn't mean anything.

One psychopath I spoke to put it like this. "Even the color-blind," he said, "know when to stop at a traffic light. You'd be surprised. I've got hidden shallows."

RIGHT TRACK?

The comparison between psychopaths and nonpsychopaths with regard to hot and cold empathy may best be illustrated through the results of brain-imaging studies.

Consider, for example, the following scenario (Case 1), first proposed by the British moral philosopher Philippa Foot:

> A trolley is running out of control down a track. In its path are five people who have been tied to the track by a mad philosopher. Fortunately, you can flick a switch that will lead the trolley down a fork in the track to safety. Unfortunately, there is a single person tied to that fork. Question: Should you hit the switch?

Most people have little trouble in deciding what to do in this scenario. Though the thought of flicking the switch is unpalatable, the utilitarian option — killing just the one person — represents the "least worst choice."

But now consider the following scenario (Case 2), proposed by the American moral philosopher Judith Jarvis Thomson:

> As before, a trolley is hurtling down a track toward five people. But this time, you are standing behind a very large stranger on a footbridge above the tracks. The only way to save the five people is to push the stranger. He will fall to a certain death — but his considerable bulk will block the trolley, saving five lives. Question: Should you push him?

Here, there is what we might call a "genuine" dilemma. Although the score in lives is precisely the same as in the first scenario (five to one), one's choice of action is far trickier. Yet why should this be?

Harvard psychologist Joshua Greene believes he has the answer, and it boils down to temperature. The reason, Greene suggests, is reflected in brain architecture — in the respective parts of the brain implicated in the resolution of each dilemma.

Case 1, he proposes, is what we might call an *impersonal* moral dilemma and involves those areas of the brain primarily responsible for reasoning and rational thought: the prefrontal cortex and posterior parietal cortex. This, if you recall, is the circuitry of *cold* empathy.

Case 2, on the other hand, is what we might call a *personal* moral dilemma and involves the emotion center of the brain. The amygdala. The circuit of *hot* empathy.

Just like you or me, psychopaths have relatively few problems with Case 1. They flick the switch and the train diverts accordingly — killing just the one person instead of five. However — and this is where it gets interesting — quite *unlike* you or me, they also experience little difficulty with Case 2. Psychopaths, without a moment's hesitation, are perfectly happy to shove the fat guy over the rails if that's what the doctor orders.

Moreover, this difference in behavior has a distinct neural signature. The pattern of brain activation in both ourselves and psychopaths is identical on the presentation of *impersonal* moral dilemmas, but radically different when it comes to *personal* moral dilemmas.

Imagine that I were to hook you up to an fMRI machine and then present you with the two dilemmas, first one, then the other. What would I observe as you went about trying to solve them?

Well, at the precise moment that the nature of the dilemma changed from impersonal to personal, I would see your amygdala and related brain circuits — your medial orbitofrontal cortex, for example — light up like Christmas trees.

The moment, in other words, that emotion kicks in.

But in psychopaths I would see nothing. The house would remain in darkness, and the passage from impersonal to personal would slip by unnoticed.

EMOTIONAL CALCULUS

Similar work in the field of face processing has been conducted by Heather Gordon and her coworkers at the Center for Cognitive Neuroscience, Dartmouth College. On an emotion recognition task (in which participants had to match a series of facial expressions presented to them on a computer screen) Gordon compared the performances of high and low scorers on the Psychopathy Personality Inventory (PPI) — a test, as we saw earlier, specifically designed to detect the presence of subclinical psychopathic traits within the general population.

Then, using fMRI, she peered into their brains to see what was going on.

What she discovered was intriguing. While, on the one hand, those scoring high on the test displayed *decreased* amygdala activity compared to those who scored lower (consistent with a deficit in "hot" emotional processing), they also exhibited *increased* activity in both the visual and dorsolateral prefrontal cortices — indicative, as Gordon and her colleagues point out, of "high-scoring participants relying on regions associated with perception and cognition to do the emotion recognition task" (see Figure 7.1).

More intriguingly still, perhaps, when it came to recognition accuracy, Gordon and her team turned up . . . nothing. Unlike with patterns of brain activity, they found no overt performance differences between those high and low on psychopathic traits — which strongly suggests that whatever strategy the psychopaths happened to be using in decoding those emotional expressions worked just fine.

Figure 7.1 — Blood oxygen level–dependent activity during the emotion recognition condition relative to a resting baseline. A: Participants who scored below the mean on the PPI. B: Participants who scored above the mean on the PPI. Sections in white indicate areas of increased brain activity.

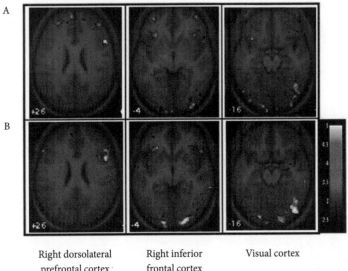

| Right dorsolateral prefrontal cortex | Right inferior frontal cortex | Visual cortex |

Simon Baron-Cohen, a psychologist at the University of Cambridge, has taken things one stage further. The "Reading the Mind in the Eyes" test (see Figure 7.2, overleaf) requires individuals to view photographs of the eye region of faces and to deduce, from this information alone, the mental state of the individual in the photograph. Give it a go yourself.

Not as easy as it sounds, is it? Most people get around two out of three (and actually, these aren't exactly the simplest examples I could have chosen either). One out of three and you're doing well here. (Answers at the bottom of the page.)*

The Reading the Mind in the Eyes test represents, as you've probably already guessed, a good index of the cold, as opposed to hot, empathy that we've just been talking about. Participants, after all, don't have to *feel* the emotion in the pictures. They simply have to *recognize* it.

Such being the case, Baron-Cohen had a brainwave. How, he wondered, in comparison with the rest of us, might psychopaths fare on the test? In

---

*Answers: top, uneasy; middle, decisive; bottom, despondent.

Figure 7.2 — The "Reading the Mind in the Eyes" test. What emotion is being conveyed in the three pictures? Select from the four options below each one.

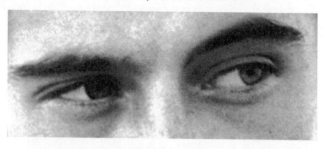

apologetic; friendly; uneasy; dispirited

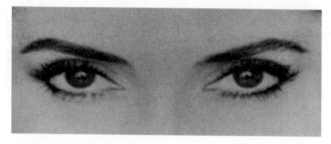

decisive; amused; aghast; bored

despondent; relieved; shy; excited

light of the brain imaging results from earlier, one would imagine that the performance of psychopaths should be pretty much indistinguishable from that of the general population as a whole.

But how would things go in the lab?

To find out, Baron-Cohen visited three London prisons and pitted nineteen psychopathic inmates against eighteen nonpsychopathic controls. He presented each individual with forty photographs of eye regions and asked them, as above, to identify the emotion in each.

Who would come out on top? Would the Starlings beat the Lecters? Or would the psychopaths hold their own?

The results were unequivocal. Exactly as predicted, Baron-Cohen found no difference whatsoever between psychopaths and nonpsychopaths. It was a tie. Plus a further indication, to go with the fMRI data, that although psychopaths may be unable to *feel* empathy, the *concept*, at least, is glacially preserved in their brains.

"I can read your brain like a subway map," Keith Barrett told me in his marble-floored hotel room overlooking New York's Fifth Avenue. "Shuffle it like a deck of cards. That's who I am. A psychological croupier. I deal the cards. Spin the wheel. Give out the chips. Then I sit back and watch what happens. Why should I feel anxious? Or bad? In fact, why should I feel *anything*? Nobody beats the house. Not in the long run. . . . You'd be surprised at how similar people are to slot machines. Know when to hold, when to nudge—and the tokens come pouring out. Emotion . . . that's for pussies."

KILLING IN THE MARKET

When one considers the circumstances under which much persuasion takes place (courtrooms, boardrooms, incident rooms, and bedrooms—to name just a few), it's perhaps not difficult to imagine how the psychopath's unique capacity to gauge, but not feel, emotion—to fly on only the one engine—may well confer an advantage. Such crystalline neurobiology as theirs is able to take, quite literally, the heat out of the moment, and allow them, in tight situations where cool, peppermint logic is at a premium, to zone in on details that the rest of us might miss. Plus, of course, to take chances—to come up with the line the rest of us might "think better" of.

"I'm the coldest son-of-a-bitch you'll ever meet," said Ted Bundy, who killed, decapitated, and screwed—in that order—thirty-five women over a four-year period.*

And he was right.

But there are times, quite clearly, when that coldness comes in handy: when instead of actually costing lives, as in Bundy's case, it can, in contrast, *save* them.

---

*The exact number of women Bundy killed is not known. He confessed to thirty murders during the period from 1974 through 1978, but estimates generally put the figure at closer to thirty-five.

The *equity premium puzzle* has long baffled financial experts. This is the tendency for large numbers of investors to invest in bonds as opposed to equities — especially during periods of stock-market decline — despite the fact that the latter, over time, have shown far better rates of return. Such conundrums as this — known as *myopic loss aversion* — have provided the impetus for a new, and somewhat timely, field of study: neuroeconomics.

Neuroeconomics focuses on the mental processes that drive financial decision making — and the primary finding so far has been that emotion is chicken. Emotion, it would seem, is so oriented toward risk aversion that even when the benefits outweigh the losses it henpecks our brains into erring on the side of caution. If Spock were a trader, consensus is beginning to form, he'd clean the rest of us out.

A 2005 study, conducted by a joint team from Stanford University, Carnegie Mellon University, and the University of Iowa, provides fascinating empirical evidence for such a claim. The study comprises a gambling game consisting of twenty rounds. At the start of the game participants are given the sum of $20 and, at the beginning of each new round, are asked whether they are prepared to risk $1 on the toss of a coin. While a loss incurs the penalty of only $1, a win reaps the grand reward of $2.50.

"Logically," says Baba Shiv, associate professor of marketing at Stanford University, "the right thing to do is to invest in every round."

But logic, as we know, doesn't always prevail.

At the beginning of the study, participants are divided into two groups: those with lesions in the emotion areas of the brain (the amygdala, the orbitofrontal cortex, and the right insular or somatosensory cortex) and those with lesions in other areas (the right or left dorsolateral sector of the prefrontal cortex). If, as neuroeconomic theory suggests, emotion really *is* responsible for risk aversion then, according to the dynamics of the game, those participants with the relevant presenting pathology (i.e., the first group) should outperform those without (i.e., the second group).

This, it turns out, is exactly what happens.

As the game unfolds, "normal" participants begin declining the opportunity to gamble, preferring, instead, to conserve their winnings. But those with problems in their brains' emotional zip codes keep going — ending the game with a significantly higher profit margin than their competitors.

"This may be the first study," says George Loewenstein, professor of economics at Carnegie Mellon, "that documents a situation in which people with brain damage make better financial decisions than normal people."

Antoine Bechara, now professor of psychology at the University of Southern California, makes an even more startling claim:

> Research needs to determine the circumstances in which emotions can be useful or disruptive, [in which they] can be a guide for human behavior. . . . The most successful stockbrokers might plausibly be termed "functional psychopaths" — individuals who on the one hand are either more adept at controlling their emotions or who, on the other, do not experience them to the same degree of intensity as others.

And Baba Shiv agrees. "Many CEOs," he says unnervingly, "and many top lawyers might also share this trait."

WIRED FOR CONFIDENCE

Shiv and Bechara's comments make sense. The arctic genius whose unblinking, remorseless neurology allows him to decouple feeling from thinking with the ease of untying a shoelace is, when the chips are down, going to leave the rest of us for dead. Sometimes quite literally.

Figure 7.3 — "Lift going up?"

Bill Gates (not someone a layperson would ordinarily label a psychopath but certainly, in business, someone whose empathy channel appears muted) was recently interviewed on television. You're a billion-dollar, multinational corporation, the presenter protested. Why do you have to crush the small guys — the two friends starting a company in their bedrooms? Why do you have to win 10–0 all the time? Gates looked at her as if she was crazy.

"I take that as a compliment," he said.

But the psychopaths' subzero brain temperature, their facility for neural cruise control, doesn't just impact on empathy. A recumbent amygdala has additional benefits, too — most notably when it comes to another of the SPICE components, confidence. Remember, not all psychopaths are behind bars — they're only the ones who engage in criminal activities and get caught. Many are law-abiding citizens out there in the workplace, excelling in high-risk occupations such as litigation, big business, the armed forces, and the media, for one simple reason: their confidence to thrive under what, to less resilient individuals, may seem like conditions of brutal, intolerable stress.

Neurosurgery is commonly regarded as one of the riskiest forms of surgery an individual can undergo. Operating in hostile environments deep within the brain, the neurosurgeon must aim for absolute precision — with margins of error less than a sniper's bullet. It's no place for the fainthearted. So who are the people who "cut it" in this profession — who patrol the remote borders between consciousness, self, and soul?

A clue comes from Andrew Thompson, a neurosurgeon for twenty-two years — who again, though not resident in psychopathy's "inner city," certainly scores high on ruthless self-belief:

> I would be being less than honest if I said that I didn't get a kick out of the challenge. Surgery is a blood sport, and playing safe all the time just isn't in my nature. . . . But one cannot allow oneself to become paralyzed by fear if something goes wrong. There's no place for panic in the heat of battle. One must strive for 100 percent concentration no matter what the eventuality. One must be remorseless, and have the utmost confidence in oneself to do one's job. . . . The brain represents the high seas of modern-day medicine, and twenty-first-century brain surgeons its pirates and buccaneers.

Thompson's comments may come as a shock to those about to undergo surgery. But they shouldn't. Sentiments like these are actually quite common among those at the business end of their professions — as the Harvard researcher Stanley Rachman discovered in a series of studies conducted back in the 1980s.

Rachman's studies are now regarded as classics, not least for his choice of participants: bomb-disposal experts. What attributes, Rachman wanted to know, did such a profession demand? What, if anything, differentiates the "great" bomb-disposal expert from the merely "good" one?

Rachman's research uncovered something interesting. Starting with a group of experienced bomb-disposal experts — those with ten years or more in the field — he began to observe a fundamental difference between those who'd been decorated for their work and those who hadn't. What's more, the nature of this difference seemed rooted in basic physiology. Rachman noticed that on assignments deserving of higher attentional resources — those, in other words, at the sharp end of the risk spectrum — the heart rates of those who had *not* been decorated remained stable.

An extraordinary finding.

But even more extraordinary was what happened to the heart rates of those who *had* been decorated. Far from remaining stable, they actually went *down*. More detailed analyses — of the impact of certain types of personality variables on cardiovascular performance — revealed the reason why. There was, it turned out, not just one factor in the mix, but two. Sure, Rachman discovered, some individuals may well, through the sheer law of averages, have ice running through their veins.

But the one defining attribute — the single most important thing that above all others seemed to make a difference — was confidence.

Confidence, of course, helps us out in all walks of life. You don't need a timer and a web of dodgy fuse wire to figure that one out. Nor a scalpel nor a cranial saw. On the golf course, in the job interview, on the trading floor, on the dance floor — it's belief in one's own ability just as much as that ability itself that often tips the balance. Just ask the victims of some of the world's best con men.

Robert Hendy-Freegard is the Hannibal Lecter of confidence-tricksters. Such are his powers of persuasion that he has, within his high-security residence, been moved to solitary confinement to prevent both his fellow prison inmates — and prison *staff* — from falling under his spell.

Over a period of almost a decade, the car-salesman-turned-con-man managed to convince his victims that he was an MI5 operative conducting an undercover campaign against the IRA. And that he could, if they so wished, recruit them, too, into the British Secret Service. Meanwhile, the money he needed "to protect state security" was winging its way out of their bank accounts.

"The most accomplished liar I have ever encountered in twenty-five years in the police" was how Scotland Yard detective Robert Brandon described Hendy-Freegard. "When he started he'd be charming; he'd listen and listen, and find any weakness in character, any vulnerability, and then he would ruthlessly exploit that. When he had control of his victims, he went to whatever ends he possibly could to take all their money and all their dignity."

Even more "impressive" is the fact that some of these victims — who included a child psychologist — were extremely well educated. Hendy-Freegard himself left school at fourteen.

So how did he do it? One of his casualties provides a clue.

"His confidence was irresistible," she recalls. "His demeanor completely infectious."

Andrew West, who spearheaded the case for the prosecution, provides another.

"I struggled to understand it," he said at the conclusion of the trial. "But he is very plausible. And even when he gave evidence he appeared very convincing."

Which just goes to show, as another psychopathic con man I spoke to put it (with only the merest hint of irony): "Anyone can *be* the part. But can you *act* it?"

THE NORMAL AND THE NUTS

In 1964, the British playwright Joe Orton penned his much-acclaimed drama *Entertaining Mr. Sloane.* In the play, the charismatic psychopath Mr. Sloane comes to live with a lonely brother and sister, and begins desultory affairs with both. Of his character, Orton wrote the following: "He has to be lethal and charming. A combination of magical black leather meanness and boyish innocence."

Which is as good a portrayal of the psychopath as any you're likely to see.

Orton's profile of the mercurial Mr. Sloane captures a facet of the psychopathic personality that is universal across many members of the species: an incongruous mélange of the normal and the nuts.

"You may have spotted them in your workplace," writes David Baines in the magazine *Canadian Business*. "They are intelligent, charismatic, attractive and socially skilled. They make great first impressions. They are spontaneous and uninhibited by rules. They are fun to hang around with — at least initially . . . [but] behind the charisma, there is no conscience."

Such incongruity is hypnotic. The allure of psychopaths is derived in no small part from their similarity to ourselves. Crossed, that is, with their obvious *dissimilarity*.

Patricia Davidson, a forty-four-year-old sales assistant from Wichita, Kansas, tells an all-too-familiar story — about how she upped sticks to Chicago, Illinois, to date a man serving life for a brutal gangland killing.

"Did he do it?" I asked her.

"Yeah," she said. "He did it. But he was never like that with me. He was a real romantic, writing me poems and stuff. He made me feel I was special. Like *I* was the one he'd been waiting for."

She wasn't. Six weeks later the relationship had petered out. Several other women had crept out of the woodwork, and Davidson had rolled back west.

Black leather meanness meets boyish innocence. It's a dangerous combination.*

COLD CALLER

The veneer of charm that makes the psychopath so beguiling also makes for excellent psychological camouflage. And when combined with demonic confidence it can be lethal.

Liam Spencer is a twenty-year-old "apprentice" A-lister — a protégé of Greg Morant, whom we met in the previous chapter. Morant, among oth-

---

*The actor Anthony Hopkins (allegedly) tells an amusing story about his "alter ego" Hannibal Lecter. Shortly after the release of *Silence of the Lambs*, Hopkins, on a visit to Wales, sneaked into the back of a small provincial cinema where the film was being shown. During the climax of the movie, where Hannibal escapes and the empty, blood-spattered elevator comes into shot, Hopkins, rather noisily, opened a bag of crisps. Pissed off — and rightly so — a woman sitting in front of him turned around. Five minutes later they were carting her out on a stretcher.

ers, is teaching him the tricks of the trade, and six months in is already tipping him for greatness.

"He's a natural," says Morant. "Ice cool, and with a predator's eye for weakness. Everyone's got an Achilles heel, it's just a matter of finding it. . . . Liam's a bit quicker than most."

Spencer is impressive. Tall, good-looking, and impeccably turned out in navy Armani pinstripe and white, open-necked shirt, he hands me back my wallet five minutes after we sit down. By which time, of course, he's already bought the drinks.

I ask him about his success with women — on a tipoff from Morant. Spencer is a man for whom getting a date on a Friday night is a sport. And he's pretty good at it, too — hardly surprising given the kinds of methods he uses.

Here's one of them, which he tells me over cocktails on the terrace.

Step 1: Check out local neighborhoods providing likely catchment
areas of single women living alone — places around hospitals,
universities, and so forth.

Step 2: Around 8 P.M. on a Friday evening, turn up unannounced at
a preselected address with a bottle of wine and a reservation for
two at one of the smarter restaurants in town.

Step 3: If a guy comes to the door, apologize, say you've got the
wrong house, and start again somewhere else.

Step 4: If a girl answers, ask if "Carmilla" (or similarly unusual
name) is at home. She won't be — because "Carmilla" doesn't
exist.

Step 5: On being told that there's no "Carmilla" at that address,
explain — with a carefully crafted, well-practiced blend of
disappointment and embarrassment — that you met Carmilla in
a bar several days ago, had asked her out to dinner this evening,
and that this was the address she'd given you. Damn — you've
been stood up!

Step 6: Stir in a pinch of humor: "I knew I was too good for her!"

Step 7: Wait for a reaction. Chances are it will be sympathetic (if
not, apologize for the intrusion and move on).

Step 8: Add a dash of opportunistic hope to your disappointment
and embarrassment. Something like: "Er, I know this might

sound crazy but, if you don't have any plans tonight (at 8 P.M.
on a Friday evening chances are she won't have) and seeing that
I'm here, I don't suppose you might like to join me . . ."
Step 9: Table for two.

Spencer's technique combines all five components of the SPICE model
into a Michelin-starred casserole of con.

Simplicity and *Perceived Self-Interest:* Goes without saying.

Incongruity: How often does a good-looking, well-dressed, funny, and
(most important of all) conveniently *available* knight in shining Armani
turn up at your door on a Friday night? With, as if all that weren't enough, a
glittering invitation to dinner?

Confidence: Could *you* do it?

Empathy: Friday night? 8 P.M.? How long does it take to get a stir-fry in
the bin? Not long at all, say several girlfriends I ask. Especially if a guy like
Spencer drops by.

But not all the time are the psychopath's motives so benign. They don't
just turn up on your doorstep to take you out for dinner. Sometimes, as
Greater Manchester Police discovered several years ago, it's to take you to
the cleaners.

INTO THE LION'S DEN

In July 2007, local police were called to a house in Manchester. There'd been
a disturbance at the address and a neighbor had picked up the phone. What
they found when they got there shocked even the most experienced officers — those who thought they'd seen it all before. They hadn't. A woman in
her thirties had been bludgeoned to death with a hammer, as had her daughter — aged eighteen — and her thirteen-year-old son. There was blood, together with an assortment of other bodily fluids, everywhere. In no time at
all the murders were front-page news.

That night, on television, detectives took a gamble. A name had emerged
of someone they wanted to talk to — and they decided to release it. He
was their man all right, no doubt about it; and given the ferocity of the attacks, and the unprecedented danger that the killer posed to the public, they
thought it best to out him. The protocol of anonymity was one thing. Innocent people's lives were another. They just couldn't take the chance.

"We're looking for Pierre Williams," said Detective Superintendent Paul Savill of Greater Manchester Police. "If you know where he is, do not, under any circumstances, approach him. He's violent and highly dangerous. And quite possibly armed. If you have any information as to his whereabouts, contact the police immediately."

A few hours later, Savill received a phone call.

"Hi," said a voice, calmly and matter-of-factly. "This is Pierre Williams. I've seen on TV that I'm wanted for a triple murder. I'm coming in."

Savill wasn't amused.

"If this is some kind of windup," he said, "I'm not in the mood."

It wasn't.

A short time later, Williams duly appeared. And Savill started to panic.

The problem was one of time. Or, more specifically, lack of it. Savill knew, as, of course, did Williams, that from the moment a suspect is taken in for questioning, the police have ninety-six hours to get their act together and press charges. If, by that time, their enquiries are inconclusive — no hard evidence has come to light — the suspect is free to go. And go, quite possibly, for good.

This was a major headache. The doors of the incident room had barely even opened. And yet here, cool as a cucumber, was the number-one suspect strolling right through them. It wasn't so much that the police had nothing to go on. They hadn't even started looking. It was, on Williams's part, a breathtakingly risky play.

Williams, needless to say, wasn't what you might call cooperative. Well aware that the trail was starting cold, he refused, as was his right, to answer any questions. Not only that, but he'd appeared on police radar before: as the cleanup man for a well-known Manchester gang — covering tracks and getting rid of forensics. This, Savill realized, added to their problems big time. With his previous history of disposing of criminal evidence, Williams, it didn't take a genius to work out, would of course have been covering his *own* tracks meticulously. Which, to use the official terminology, was a king-sized pain in the arse.

Eventually, as it turned out, Savill *did* get his man. A latent footprint — invisible to the naked eye — at the scene of the murders in Manchester proved a positive match with a footprint at Williams's apartment. In Birmingham, one hundred miles away. The judge sent him down for life.

But it was a close-run thing. Just three hours remained on the clock when news of the breakthrough arrived.

Savill breathed a massive sigh of relief.

"Williams coming forward immediately of his own volition was totally unexpected," he admitted later. "No one could have predicted that. Right from the start he had us on the back foot. But none of us would've forgiven ourselves if we'd let him slip through the net. We knew he was our man and it was persistence and good old-fashioned police work that finally won the day. But there's no doubt about it — we went right down to the wire on that one."

Even with a squad member down (it was in the police's interest to detain their man *not* release him), SPICE is a force to be reckoned with. Through the audacious deployment of the four remaining factors — simplicity, incongruity, confidence, and empathy — Williams, quite literally, almost got away with murder.

MIND THE CRAP

When Hannibal escapes in *Silence of the Lambs,* Starling is convinced he won't come after *her.* "To him," she surmises, "that would be rude." And she's right. But not all psychopaths are as accommodating as Lecter. Their facility for flouting social norms, for doing the unexpected, can, as we've just seen with Pierre Williams, often be electrifying. And significantly enhances their capacity to charm and persuade.

In his book *The Stuff of Thought,* the Harvard psychologist Steven Pinker talks about *implicatures.* An implicature is a linguistic device that allows us to say what we mean by saying stuff that we . . . well, *don't* mean.

A classic example is often heard at the dinner table. Imagine that you're sitting down with a group of strangers and want someone to pass you the salt and pepper. You turn to the person next to you and say . . . what, exactly? Well, the chances are you *won't* ask them to pass the salt and pepper. You won't say what you *mean.* What you'll probably say instead is something along the lines of: "*Would you mind* passing the salt and pepper?" or "Can you see the salt and pepper anywhere?" Anything, in other words, but a short, sharp, simple, and straightforward "Pass the salt and pepper!"

Implicatures exist, Pinker argues, because they enable us to save face. They inoculate against effrontery. The request "Pass the salt and pepper" could, on the one hand, be interpreted as a directive. Less of a request, and more of a barefaced challenge. On the other hand, however, "Can you see the salt and pepper anywhere?" lets us off the hook. We all *know* what it means (JUST GIVE ME THE BLOODY SALT AND PEPPER!) but some-how—because intention is *implied* rather than asserted—it doesn't seem as bad.

When I heard about implicatures I went over to Harvard to have a chat with Pinker. Split-second persuasion, it seemed, didn't fit the bill, or so I thought at the outset. Take the following, for instance.

A husband and wife are involved in a shouting match in their local bar. It's the August bank holiday, and both of them are pissed. The pub is full of regulars, and the argument has been raging on for a good quarter of an hour.

"You never tell me the truth!" yells the husband. "That's always been the problem with you. You're never honest with me. Why don't you just stop all the bullshit and give it to me straight?"

"Yeah," says the barman. "Why don't you just drop the bullshit and tell the short, fat, bald, stingy twat the truth from now on?"

See what I mean? Statements like these aren't exactly *teeming* with lines to read between, are they? They don't leave *too* much to the imagination.

But Pinker viewed it differently.

> We've basically developed all these linguistic strategies to protect our-selves, and they work. But they're also a pain. So when somebody de-cides to break the rules and just say it as it is—to cut through all the crap—it can, depending on the context, be kind of refreshing. It can be a relief. It's the basis of a lot of humor, for example. And because our notions of politeness are shared—I know that you know that I know you're breaking the rules—there's still a kind of safety net there. . . . It seems to me that the power of split-second persuasion lies in its fresh-ness. It's basically a cutting-through-the-crap kind of influence.

I liked where Pinker was coming from. Ironically, he seemed to be say-ing, it's precisely *because* we have implicatures that SPICE can work its magic. Because it's *less* of the same. We're all so busy going around trying not to offend each other that when someone comes along and raises the fin-ger to linguistic convention our brains breathe a sigh of relief.

Suddenly my thoughts turned to psychopaths.

Little wonder they were the kings of persuasion. Irrespective of what Starling might have thought of Hannibal (and, let's face it, he wasn't *that* po-lite), this was exactly what they were good at. Their hardwired impulsivity and electrifying charisma made their use of incongruity second nature. Not only that, but they were good at something else. Something directly related to what Pinker was saying about SPICE.

When the chips are down and they stand to gain out of a situation, psychopaths zero in. They're second to none at "taking care of business." At focusing on ends rather than means. Or, as you may choose to say if you're one of the world's greatest linguists, "cutting through the crap."

JUST REWARD

Imagine that I were to present you with a set of sixty-four cards, displayed one after the other on a computer screen, each bearing a two-digit number between 1 and 99. There are eight of these two-digit numbers — and each will appear a total of eight different times during the course of the presentation. Your task is simple.

You must decide which of these numbers to respond to by pressing the X key on the keyboard, and which of them to respond to by pressing Y.

The only catch is that each time you get it wrong you'll receive a painful electric shock.

How do you think you'd get on?

A few years ago, the psychologist Adrian Raine and his colleagues at the University of Southern California in Los Angeles performed an experiment to find out. What they turned up was astonishing.

If you're like most people, you'll pick up the "rule" pretty quickly (e.g., X = odd numbers; Y = even numbers). Once you've had one jolt, you won't want another in a hurry.

Except, that is, if you're a psychopath. With these guys, something rather odd happens.

On tasks such as these — called *passive avoidance learning* tasks — psychopaths, time and again, commit significantly more errors than the rest of us. The threat of impending punishment, the prospect of danger or discomfort, just doesn't seem to bother them in the same way it would you or me.

They just couldn't, it would seem, care less.

Findings such as these may seem to indicate that psychopaths, on the face of it, just "aren't interested." That their singular lack of emotion quite simply "tunes them out." And this, *on the face of it,* sounds reasonable enough.

But now let's imagine a slightly different scenario. This time, let's imagine that we have exactly the same deal — the cards, the numbers, the shocks — but on this occasion, should you get it right, you don't just avoid the punishment, you also get a reward: $5 a hit.

Reckon that would make any difference? That you'd figure out the rule even quicker? Most people reckon no: the electrodes do just fine. But in situations like these, fortunes change dramatically.

Psychopaths, as if by magic, actually perform *better* than the rest of us.

In contrast to those scenarios in which the emphasis is on avoidance of the negative, when there is actually something to gain from the enterprise they pick up the rule a lot faster. Appeal to a psychopath's self-interest and nothing much gets in the way.

HOW TO WIN FRIENDS — AND LOCK THEM UP

The ability of the psychopath to flatline under pressure, to deliver the goods when the rest of us may drop them and run, has not been lost on the movie business.

The Dirty Dozen — set in World War II — features a band of twelve desperados on a do-or-die mission to destroy a French château full of high-ranking German officers. Ultimately, the mission proves successful — although only one of the twelve survives. But beneath this saga of redemption lies an interesting question: Why was this band of renegades picked for the job in the first place? Why entrust a mission of such paramount strategic importance to a cohort of rapists and murderers? Might the Hollywood dream factory have actually been tapping into something a little more "real" here? Might the eagle have a better chance of landing if the roughnecks are calling it in? There's evidence to suggest that it might.

In Britain, the Conspicuous Gallantry Cross is bestowed upon its recipients "in recognition of an act or acts of conspicuous gallantry during active operations against the enemy." Since its introduction in 1993 it's been awarded on a mere thirty-seven occasions. Here's an extract describing one such occasion, taken from the *Independent*:

> It was during fierce fighting in the caves of Tora Bora, a Taliban stronghold, as part of the hunt for Osama bin Laden in 2001 that SAS Regimental Sergeant Major Bob Jones [his name has been changed] took

on the enemy armed only with his commando knife, despite having been seriously wounded. He was hit at least twice by enemy fire, yet he somehow managed to get back to his feet and continue fighting, before resorting to his knife as the conflict descended into savage hand-to-hand contact. . . . Officials described his "outstanding leadership in drawing his knife and charging the enemy, inspiring those around him when ammunition was running low and the outcome of the battle was in doubt."*

One sees similar focus among the world's top con artists. Greg Morant, by his own admission, prepares like an "Olympic athlete" when there's something big in the offing.

> I'll find out everything I can [about the person]. From how they do business to what they do on the weekend. Top athletes study videos of their opponents, dissect their game. I do the same: gather information, start building up a picture of who I'm dealing with. It's not rocket science. The more you know about someone, the more the odds are stacked in your favor. . . .
>
> I'm like a surveyor. But instead of buildings I do minds. I go round and round with a fine-toothed comb. Looking for doors, for that secret way in. And you find it eventually. There's always a way in. Round the back. Out of sight. Hell, sometimes you just walk straight in through the front!
>
> It's what everyone should do if they're serious about getting ahead. I don't mean to sound disrespectful, but all these books you read about how to persuade and influence people? Most of them are full of shit. You can talk all you want about psychology. But the trick is to do your homework. To start a fire you've got to have something to burn. And not everything burns, right?

If psychopaths can "make" out of a situation, if there's any kind of reward on offer, they go for it. And go for it big time—irrespective of risk

---

*October 1, 2006. This is not in any way to imply that "Bob Jones" is a psychopath or in any way an "undesirable"—so don't come after me, Bob! The point I am making here is simply that there are certain psychopathic traits, in this case focus and immediate disregard for one's own personal well-being, that can, under the right circumstances, predispose one to greatness.

or possible negative consequences. Not only do they keep their composure in the presence of threat or adversity, they actually go one better. They become, in the shadow of such presentiment, laserlike in their ability to "do whatever it takes."

An example of such focus — in conjunction with the use of SPICE — comes from a friend of mine. Paul and I were at university together. Though he had little in common with Hannibal Lecter (during the entire time I knew him he just about managed a parking ticket), Paul was, is, a psychopath. I know because I tested him. But also, of course, there were the usual telltale signs. Suave, intelligent, ruthless, and confident — the most striking thing about Paul, the characteristic that everyone knew him for, was his devastating ability to persuade. His ability to engender trust. It was, quite literally, as if he had some secret piece of software buried deep within his brain that enabled him to hack into the innermost thought patterns of others. And then, once he had gained access, to do anything he wanted. If Paul didn't know your emotional password already, it would take him less than five minutes to decipher it. He was (and no doubt still is) one of the most gifted psychocryptographers I've ever known.

The last time I saw Paul was around seven years ago, and he'd lost none of his ability to turn a situation to his advantage. Picture the scene: A crowded train carriage in London, two builders covered in dirt and paint, and Paul in a neatly pressed pinstripe suit sitting opposite them. It's been raining most of the day and the builders — who've obviously been working outside — are wet through. They start to give Paul a hard time.

BUILDER NO. 1: You've got an easy number, haven't you? Sitting
    there in your suit and tie. A real day's work would kill you.
PAUL: Whose job would you rather have — yours or mine?
BUILDER NO. 1: You're having a laugh, aren't you? I wouldn't be you
    for a second!
PAUL: Fine. Then what are you complaining about?
BUILDER NO. 2: Clever bastard, eh? Well, let me tell you this. If he
    doesn't want it, I'll have it!
PAUL: Fine. Then what are you complaining about? You're just
    jealous.

One of Paul's girlfriends (he had many) once told me a story that epitomizes his improvisational genius. One night, as they lay in bed, they were awoken in the small hours by a burglar. It was dark, but Paul could just about make out the profile of the intruder—barely a couple of meters away—hovering over his Powerbook on the dresser. Whereas most people would pretend to be asleep—or, in blind panic, do something they'd later regret—Paul remained calm and focused.

"Look," he said through the darkness, in an even, matter-of-fact tone, "I don't want to get into a fight with you or anything—even if it means having to put down this semiautomatic I've got trained on you under the duvet! I've broken into a few houses myself in my time [this was a lie] and I firmly believe that what goes around comes around. So I suppose, when it comes down to it, I'm really not that bothered if you take the Powerbook.

"In fact, I'll cut you a deal. If you just let me copy a couple of things off the desktop, I won't even report it missing. For a start I can't see your face. And if you're smart you'll have gloves on anyway. So at the end of the day there's no real point in me going to the police, is there? How about it?"

Frozen in terror, Paul's girlfriend lay beside him as the intruder, silhouetted against a nightlight on the landing, thought about his offer. Amazingly, after what seemed like a lifetime, he decided to go for it: Paul's magic had worked again. But this was just the beginning. Having now taken control of the situation, Paul *really* began to motor.

First, he suggested to the burglar that he take a step outside for a moment while he began downloading. The light from the computer screen might illuminate his face—and it would only prey on his mind if he allowed himself to be identified. He did as he was told. Next, as he sat at the dresser hoovering up his desktop, Paul began to chat to him. He started to make up details of the houses he'd broken into, and of how the abuse he'd suffered from his stepfather when he was a child had driven him to crime. (In fact, Paul had had a very happy upbringing.) Lo and behold, the intruder began to talk about his own traumatic childhood, and the two of them struck up a conversation. They started to get along.

When Paul had finished downloading, he cut the intruder a second deal. Why didn't they continue the conversation downstairs in the kitchen over a beer? Though the circumstances were, to say the least, unusual, Paul felt that fate had brought them together. They seemed to have a lot in common. And besides, he was having trouble sleeping anyway. Again, the intruder swallowed it. As an afterthought (the burglar didn't actually ask him to do this—this was just Paul getting into his stride), he threw a balaclava out through the bedroom door and told him to put it on. That way he wouldn't be recognized. Then the two of them went downstairs.

Sure enough, as she hovered on the landing with a bath towel draped around her, Paul's girlfriend heard the sound of the fridge door opening and two ring pulls being popped down below. Then, a little while later, two more. Eventually, despite Paul's protestations, the burglar even took off his balaclava. He'd begun to feel at home.

The pair of them chatted for well over an hour. If you'd just been passing and didn't know who they were, you'd have bet your bottom dollar they'd known each other for years. When, finally, the burglar called it a night, a dozen or so cans lay crumpled up on the table and the first signs of light were creeping in through the curtains.

But before he went, Paul had an idea. Maybe the two of them should team up together. Being a postman (he was, in fact, in finance), he had insider knowledge of when people in the area were going away on holiday. Intelligence like that was priceless, he pointed out. Scarcely able to believe it, the burglar gave Paul his address and telephone number. They shook hands. Paul said that he'd call round in a day or two to talk business. And the burglar said "Great!" — he'd get some more beers in.

Paul also insisted that the burglar — despite the fact he no longer wanted it — still take the Powerbook. "A deal's a deal," he said.

Next day, of course, the burglar *did* receive a visit. But it wasn't from Paul. Several members of the local police managed to retrieve not only his Powerbook but a whole load of other items that had been reported missing over the previous few months.

Paul got a personal letter of thanks from the superintendent, plus a commendation.

Simplicity. Perceived self-interest. Incongruity. Confidence. Empathy. The dark persuasive wizardry of the psychopath.

SUMMARY

Within society, there have always been elites. There's an elite in sport, in intelligence, and in social class. But there's evidence to suggest that there are also elite *persuaders*. And that quite a few of them are psychopaths.

Most people think of psychopaths as monsters. As rapists, serial killers, or terrorists. And they're right. Many rapists, serial killers, and terrorists *are* psychopaths. Yet, contrary to popular belief, a great many psychopaths don't even break the law. Instead, they head up multinational corporations, conduct high-wire brain surgery, storm embassies and airplanes in balaclavas and respirators, and invest our money in lucrative — if volatile — markets.

This coolness under pressure, this superior neural aircon, equips the psychopath perfectly for persuasion. A dysfunctional amygdala — the part of the brain that processes and experiences emotion — and the attendant absence of fear that almost invariably accompanies it, allows those who exhibit such anomalies to take chances. To zone in on outcomes unimpeded by convention. To go for shots that to the rest of us seem unthinkable.

When you're as cold as ice and have the confidence to match, a hole in one is always on the cards.

In our final chapter, we continue our exploration of the frontiers of influence — moving from the ultimate *persuader* to the ultimate in *persuasion*.

Split-second persuaders like Paul may well be the best in the business at decoding our brains' security systems. But are there some combinations that even they can't crack?

Does every lock of influence have its key?

Or does persuasion — even SPICE — have its limits?

8

Horizons of Influence

A man is walking through the streets of Belfast one night when he feels a gun against his head. "Protestant or Catholic?" says a voice. Thinking quickly, the man replies, "Jew." "Then I must be the luckiest Arab in Ireland," says the voice.

At the height of the Second World War, a disguised Winston Churchill is on his way to a secret underground location to broadcast a speech to the nation. His aide hails a taxi, and gives the driver the address.

"I'm sorry," replies the driver. "I'm on my way home. The prime minister's on the radio in five minutes and I don't want to miss him."

Impressed by the man's loyalty, Churchill whispers to his aide to hand him a £10 tip. "Fuck the prime minister!" says the driver. "Where do you want to go?"

MIRROR MIRROR

H. L. Mencken, the American humorist, once said that for every problem there is a solution which is simple, clean, and *wrong*. But consider, for a moment, the flip side of such a statement: that for every problem there is indeed a solution, but one that's simple, clean, and *right*. That there exists, in some swanky Platonic zip code — uncontaminated by ego or misunderstanding — a master key to persuasion in perfect, pristine form. How feasible is such a notion — that any mind, at any time, *really can* be changed? If it *is* true, what kind of key might it be? And how do we go about finding it?

Several years ago, when I first started thinking about this problem, I rang up Bob Cialdini. Cialdini, whom we've met a number of times already in this book, is professor of psychology and marketing at Arizona State University, and one of the world's leading experts on persuasion. I put it to him that persuasion, in theory at least, has no limits. He agreed.

"When you look at what happened in Jonestown," he responded, "how the Reverend Jim Jones managed to persuade those nine hundred people to take their own lives . . . you're talking pretty extreme mind control there. Maybe over the short term persuasion has its limits, but over the long term . . . I'm not so sure."

For quite some time after talking to Bob Cialdini I was pretty convinced he was right. I was, in fact, prepared to go one step further. The examples of extraordinary turnarounds that my own research had uncovered appeared, as we've seen, to allude to something deeper. That even over the *short* term the power of persuasion was infinite. That the solutions were out there — it was just a matter of finding them.

But then something happened that completely changed all that. I came face to face with the Mirror Man.

I first ran into the Mirror Man in the spring of 2008. Max Coltheart, professor of psychology at Macquarie University, had mentioned him at a conference. I was intrigued. I e-mailed Coltheart several days later and asked if I could visit. Fine, he said, just don't expect any miracles. I won't, I lied. And hopped on a plane to Australia.

The Mirror Man is one of the strangest case studies ever to emerge from the annals of neuropsychology. (And there have certainly been some strange ones over the years.) The encounter took place in Sydney, at the University of Macquarie's Centre for Cognitive Science, where Coltheart has founded the Belief Formation Programme: a project designed to unravel the causes of delusional ideation and develop a model of belief acquisition and rejection.

He's certainly not short on raw material.

Up until now, the program has hosted a bewildering array of notional malapropisms ranging from the more common type of delusions usually found in schizophrenia (delusions of *persecution* — people are out to get you; delusions of *reference* — private conversation and background social "noise" is targeted specifically at you; and delusions of *control* — alien forces are controlling, or intercepting, your thought patterns) to an even *more* bizarre category of cognitive misapprehension: *monothematic* delusions.

Included in this latter taxonomy is the *Capgras delusion* (the belief that someone emotionally close to you — most typically your partner — is not who he or she appears to be but, rather, an identical-looking impostor); the

Cotard delusion (the belief that you are dead); and the *Fregoli delusion* (the belief that you are being stalked by a group of people who are disguising themselves to conceal their true identities).

Then there's the daddy of them all: the *mirrored-self misidentification* delusion.

The Mirror Man, whom I'll call George, is in his mideighties. He's friendly, married with two children, and, after a successful career in business, still keeps his hand in running an advertising firm with his wife. God, I think to myself, when I see him. This guy is just so normal. Can it really be true — all these things I've heard about him? I soon find out that it is.

One of Coltheart's coresearchers, Nora Breen, presses the button on a TV remote and we enter a room with a mirror. George stands in front of the mirror and Nora asks him: "Who do you see in the mirror, George?"

George sounds apprehensive.

"It's him," he says.

"Who?" asks Nora.

"*Him*," says George. "The guy who follows me around. The guy who dresses like me. And looks like me. And does everything at exactly the same time as me."

Nora shuffles into the picture.

"Who do you see now?" she asks.

"That's you," says George. "And him."

I'm gobsmacked.

Ask him to explain how you're standing next to him *in front* of the mirror, I will her, and yet the guy who's staring back is someone else entirely.

She does.

George shakes his head.

"Look," he says. "I know it sounds crazy but that's just the way it is. I wish I *could* believe it was me in there. But I can't. It's some other guy. He looks like me. Acts like me. Does everything at exactly the same time as me. But it's just *not me*! It's *him*."

"Thanks, Nora," I say, and pour myself some coffee.

We decide to leave it there.

CRISIS OF CONVICTION

My encounter with the Mirror Man gave me quite a bit to think about. What I'd witnessed in Max Coltheart's lab wasn't just George having a bad day. It was George having a *good* day. He was, in fact, quite a star at the Belief Formation Programme. Staff there had tried pretty much everything to help him but had, quite clearly, run into a brick wall. George's conviction that the man in the mirror was an impostor — and not himself — remained as rock solid as ever. And things, Coltheart surmised, weren't going to get any better. No matter what they threw at him.

Suddenly, my thoughts returned to Jonestown. There was, it occurred to me, a whopping great paradox here. On the one hand, the Reverend Jim Jones could persuade nine hundred people to guzzle away their lives — not just their own but those of their children, too — around a big sloshing tub of arsenic-laced Kool-Aid. But on the other, some of the world's top psychologists were having trouble with a man and a mirror: convincing him that it really was *him* in the reflection, and not — as he insisted — some dastardly, double-dealing alter ego.

The implications were intriguing. Either there was something a little bit special about people like Jones, and we needed to find some in Sydney. Or there was something going on with the anatomy of belief itself. Some spectrum of strength along which all conviction lies. Unshakeable at one end, ephemeral at the other; with a sliding scale of influence, of openness to persuasion, in between.

YES, WE CAN

In the summer of 2008, not long before Barack Obama's presidential nomination, Ray Friedman, professor of management at Vanderbilt University, and two coresearchers sampled twenty questions from the verbal section of the Graduate Record Examination and cobbled them into a test. They administered the test to two groups of Americans — African Americans on the one hand, Caucasians on the other — and averaged up the scores for each one. Several months later, when the election was done and dusted and Obama had been sworn in, they gave out the test again. To exactly the same groups. And totted up the averages like before.

What Friedman and his coworkers were hoping to find was the opposite of what their predecessors had found a decade or so earlier. Back in the 1990s, groups of students with identical SAT scores had taken a similar test at Stanford. There, researchers had discovered that African Americans performed significantly worse on the GRE-style problems when asked, at the outset, to tick a box that indicated their ethnic background. The reason for the disparity was clear. Ticking the box had done more than provide demographics. It had, for the African Americans, activated the racial stereotype of academic inferiority. It had told them, *No, you can't.*

A generation on, and Friedman was after the leveler.

"Obama is obviously inspirational," he says. "But we wondered whether he would contribute to an improvement in something as important as black test-taking."

Incredibly, it turned out he could.

Analysis revealed that *prior* to Obama's nomination, whites, on average, scored around 12 out of 20 on the questions — compared with around 8.5 for blacks. But when the tests were administered *subsequently* — immediately *after* Obama's nomination acceptance speech, and then again, after his inauguration — it was a different story entirely. On both occasions, the performance of the African Americans improved significantly.

The writing was on the wall. Nobody had suddenly gotten brighter here. We're talking, after all, about only a matter of months. It was simply a case of harnessing the power of conviction.

Of believing that, *Yes, we can.*

TWO MINDS

Friedman's results have currently still to be replicated. And, in fairness, the bungeelike ease with which the performance gap contracted exceeded even his own expectations. But there's evidence to suggest that he and his co-workers are on the right track. And that Henry Ford might well have had a point: if you believe you can, or believe you can't — either way you're right.

Jeff Stone, a psychologist at the University of Arizona, has demonstrated effects similar to Friedman's in sport. In a study which pitted black against white on the golf course (we touched upon this briefly in Chapter 3), Stone has shown that when golf is framed as a test of *athletic prowess*, it's the black players, on average, who post the better scorecards. But guess what? When the game is portrayed as a measure of *strategic ability*, and the athletic component swept quietly under the carpet, fortunes change. It's the white guys who stake out the fairways — while the black guys strike out in the rough.

Then there's Margaret Shih and her Asian women math students. Asian women, if you recall, perform *better* at math when they think of themselves as "Asian" (i.e., when the *racial* stereotype is activated) — and *worse* when the spotlight suddenly switches to *gender* (i.e., when they think of them-selves as "women").

And such observations have nothing to do with effort. It's not that the women try harder all of a sudden when magically, by decree of random se-lection, they happen to "turn Asian." Far from it. Like Friedman's students, and the golfers tested by Stone, they are, instead, *persuaded* to do better. Not, perhaps, in the traditional sense of the word — through incentives, re-wards, or the usual methods of everyday social influence — but by spiking their brains with confidence. By switching their mind-sets from one com-ponent of endogenous self-identity, to another.

Cognitive psychologist Carol Dweck, at Stanford, has done some inter-esting work on mind-sets. In support of the idea that some beliefs are harder to shift than others (and some individuals harder to persuade), Dweck has identified two different styles of thinking: ways of relating to the world that can, according to her research, ultimately predispose us to success or failure in life.

Mind-sets, according to Dweck, have one of two signs up in the win-dow: open or closed. Those with the closed sign Dweck calls "fixed." This mind-set, she claims, belongs to people who "do things their way" — who are wary of exceeding their comfort zones, who see effort as negative, and who are averse to being stretched. Those displaying the open sign, in con-trast, Dweck refers to as "growth" mind-sets. People with this kind of mind-set tend generally to be more flexible — more amenable to learning, and open to the prospect of challenge. And prefer, unlike those with a fixed mind-set, to assimilate the viewpoints of others.

Dweck has shown that it's possible to manipulate these mind-sets. And, moreover, that each is accompanied by a distinct neural signature. In one study, students were split into two groups. One group was presented with arguments that supported a "fixed" mind-set (e.g., "Your intelligence is something very basic that cannot change much"), while the other group was presented with arguments in favor of a "growth" mind-set (e.g., "No matter how intelligent you are, you can always improve"). Then, both groups com-pleted a difficult reading comprehension task (on which they performed poorly) and were asked, after receiving feedback, whether they'd like to check out the answers of some of the other participants in the study: ei-ther those who'd performed better than they did, or those who'd performed worse.

Exactly as Dweck predicted, the fault line ran right down the middle. The students who were exposed to the fixed mind-set literature went straight for those who'd done *worse:* it boosted their self-esteem. On the other hand, however, those presented with the growth mind-set arguments gravitated the other way. *They* chose the answers of those who'd done *better* — compared themselves *upward,* in other words — to hoover up strategies that might help them out in the future.

But that wasn't all. Alongside this chasm of comparison a secondary fault line materialized — inside participants' heads. In a follow-up experiment using EEG, Dweck monitored patterns of cortical brain activity as the students took part in a general knowledge quiz. The experiment consisted of two parts. The first part began after participants provided their answers: a second and a half after each response was entered, a program revealed whether the students had got it right. Part II kicked in a second and a half after *that,* when up flashed the actual solution.

The data mapped on to the behavioral findings perfectly.

Exactly as the results of the previous study suggested, Dweck found that students who presented with closed, or fixed-style mind-sets entered, as expected, a heightened state of vigilance during the *initial* phase of the experiment (while they waited to see if they'd answered the questions correctly).

But then their brains zoned out. They simply cut and ran.

The students with the growth mind-sets, on the other hand, exhibited a different pattern entirely. Sure, during the first phase of the experiment — while they waited for either the thumbs up or thumbs down — their brains, just like those of their fixed mind-set counterparts, "switched on." But then (in contrast to their fixed mind-set counterparts) instead of switching off once the cat was out of the bag, they continued to twitter — sustaining the neural chit-chat right the way through the next second and a half, while they awaited confirmation of what the answers actually were.

Some of us, it seems, are genuinely open to influence. Others just want to be "right."

JUST CAN'T HELP BELIEVING?

The results of Carol Dweck's research, and the work of Ray Friedman and Jeff Stone among others, jell rather nicely with a quantitative take on persuasion. They support the observation that some people — extreme fundamentalists, for instance — have a mind-set so fixed, have neurons that fuse together so solidly on impact, that they can, at times, be almost impossible to influence. And that others just go with the flow.

Some of this may well be innate. Take a look inside any classroom or playground and you'll see both sides of the coin: kids who are traumatized by even the slightest criticism or challenge, and kids who take it all in their stride. (Want to see how easily persuaded *you* are? Then why not take the test on p. 252?) On the other hand, however, we all have our moments. Our islets of fanaticism on which only the in-crowd are welcome. This suggests that environment is also important — shaping, over time, not only our attitudes in general but also, over the short term, dictating those values most salient in our lives. (Relatives, for example, of those killed in Iraq or Afghanistan are likely to have stronger views on British or U.S. foreign policy than those, perhaps, with less personal involvement.)

But it also hints at something a little deeper — a general, founding principle of how our brains make up our minds. If belief and emotion are so deeply intertwined, could it be that our brains are somewhat less discerning than we think? That they leap *before* they look? That *first* they believe and *then* appraise and consider? And that the views we espouse are not the views that we've reasoned ourselves *toward* — but instead are the views we've been unable to reason *away* from?

Though such a notion seems crazy, there's evidence to suggest that it's true.* And that the feeling we get when we're served up new information — of chewing it over and deciding, morsel by morsel, whether or not to swallow it — is actually an illusion.

Harvard psychologist Dan Gilbert and his colleagues conducted a study in which participants were told about a robbery. Volunteers were divided into two groups. One group read statements which *exacerbated* the severity of the crime (e.g., "Kevin threatened to sexually assault the clerk"), while the other group read the opposite: statements that *extenuated* the raid (e.g., "Tom apologized to the clerk for having to rob the store").

Somewhat unusually for a psychology experiment, the researchers came clean from the start. Right off the bat, both groups were told that the character descriptions were bogus. Yet while the participants were reading about

---

*The idea was first proposed by the seventeenth-century Dutch rationalist philosopher Benedict de Spinoza.

the robbers, some of them were interrupted: the researchers assigned them a counting task. Such distraction, Gilbert proposed (if first we believe and then "unbelieve"), should interfere with the "unbelieving" part of the equation. It should, in those crucial few milliseconds during which the brain, having taken in information and "believed" it, decides whether or not to *continue* believing it, divert its attention to a completely different task, much like the "removals men" in Chapter 3 diverted the attention of the hotel receptionist at the wedding — and then carted off the presents. And should, even though the participants were told quite clearly that the character references were false, in fact make them think they were true.

This, it turned out, is exactly what happened. When, at the conclusion of the study, the participants were asked to pass sentence on the robbers, some interesting verdicts came in.

Mr. Nice, on average, got 5.8 years; Mr. Nasty, 11.2.*

And this, remember, was in spite of the fact that the participants — right from the word go — had been told quite clearly that the character descriptions were *false*.

Sometimes you can't unbelieve *everything* you read.

BELIEF IMMUNODEFICIENCY

The implications of Gilbert's study definitely take some getting used to. On the other hand, however, certain things fall into place. We can see, all of a sudden, why empathy and perceived self-interest are so important in persuasion. If, by using the right combination of words, by invoking the right linguistic force field, we can frame something in such a way that makes whoever we're talking to *want* to believe it, then we're halfway there already. Because right from the gun, they actually believe it anyway: at least, that is, for the first few hundred milliseconds! Our job as persuader is easier than we think. It's not to get others *believing* what we say. It's to stop them *unbelieving*.

Then, of course, there's the role of incongruity. Recall, from Chapter 6, how customers were more likely to purchase Christmas cards from a door-

---

*No such disparity was found with those participants who were *not* distracted (six years for Mr. Nice vs. seven years for Mr. Nasty). These guys were able to "unbelieve" the false characterizations — leaving no residual difference between the nature of the two crimes.

to-door salesman if he unexpectedly quoted the price in cents rather than dollars? And how visitors to an outdoor market bought more cupcakes from a confectioner if he referred to them as "half-cakes"? There was, if you remember, a catch. The scam only worked if the vendor, immediately after his unusual opening gambit, slipped in a caveat: in the case of the Christmas cards, "It's a bargain!" and in the case of the cupcakes, "They're delicious!"

It doesn't take a genius to work out what's going on here. It's simple: the "Gilbert effect" in reverse. Customers are offered the standard one-line sales patter — "It's a bargain!" or "They're delicious!" — but are so bamboozled by the nonsense that's just *preceded* it that they "forget to unbelieve." They are so preoccupied with getting their heads around half-cakes, or however-many-thousand cents they have to fork out for the Christmas cards, that their brains' central locking systems conveniently fail to activate — and the doors are left wide open.

The conclusion seems pretty clear. Shut down production of the brain's unbelief antibodies — for sufficient length of time as to enable whatever virus of information we wish to introduce to multiply and gain a toehold — and then persuasion should have no limits. The problem, of course, is knocking out the system.

PERSUASION UNDER PRESSURE

I got my own, very personal taste of the belief immunodeficiency virus while doing a pilot for a TV show about persuasion. The episode centered on persuasion in a military setting. What, I wanted to know, were the characteristics of a good interrogator? Could any of us do it? Or was there, as with everything, a spectrum of natural talent? Iconic images of interrogators in popular culture — Laurence Olivier in *Marathon Man,* for instance — suggest that it's badness rather than brilliance that tips the balance. Yet research conducted in both military and forensic settings sheds a radically different light on the subject. Rather than resorting to violence, the world's most sophisticated interrogators have much in common with the world's most sophisticated con artists. They infiltrate rather than invade. Work our minds rather than our molars. And possess an intuitive grasp of primeval "street psychology."

To find out what my own limits were, I was pitted, with a few reservations on *both* sides, against the professionals: Ivy League expert on persuasion trades the lawns and libraries of Cambridge University for a head-to-head contest with Special Forces mind-hacks. I was to be given three pieces of information which I was to endeavor to conceal from my "captors" — who, in turn, would deploy a lethal combination of physical and psychological techniques to try to get them out of me.

It seemed like a good idea — until I met one of the interrogators.

"What level of violence should I expect?" I asked Dave, as we sat sipping lattes in Starbucks.

He smiled.

"It's not the violence that'll break you," he told me. "It's the *threat* of violence. That carcinogenic thought process that something terrible is going to happen and that it's just around the corner." "Are you sure you should be telling me that?" I joked. "It makes no difference at all," he replied. "Even though you know in advance that we're not going to kill you, it won't be of any use. It's what's in here [he tapped his head] that's going to finish you. Sure, you might believe right now that we're not going to kill you. But once we get rolling, it won't take much for the boys to convince you otherwise."

To be honest, I was skeptical. But then Dave gave me an example of the kind of thing that happens in Special Forces selection — the kind of thing he had in store for *me*.

> Typically by this stage the candidate's exhausted. . . . Then the last thing he sees before we place the hood over his head is the two-ton truck. We lie him down on the ground, and as he lies there he hears the sound of the truck getting closer. After thirty seconds or so, it's right there on top of him — the engine just inches away from his ear. We give it a good rev and then the driver jumps out. He slams the door and walks away. The engine's still running. A little while later, from somewhere in the distance, someone asks if the handbrake's on. At this point, one of the team — who unbeknown to the guy in the hood has been there all the time — gently starts to roll a spare tire onto his temple as he lies on the ground. You know, by hand. Gradually, he increases the pressure. Another member of the team revs the truck up a bit so it seems like it might be moving. After a few seconds of that, we take the tire away and remove the hood. Then we lay into him.
>
> "Tell us your fucking name." It's not unusual for people to throw in the towel at that point.

When it came to it, *my* "moment of truth" wasn't entirely dissimilar. Chained naked to the floor of a shadowy, disused warehouse, I watched —

seemingly in slow motion — as a huge forklift truck dangled a pallet of reinforced concrete ten or so meters above my head, and then gradually lowered it so that the rough, splintery base exerted a gentle pressure on my chest. It remained there for about fifteen seconds before I heard the operator yell above the scream of the hydraulics: "Jim, the mechanism's jammed. I can't get it to move."

Dave was right. In hindsight, in the safety of the debriefing room, it soon became clear that I'd been in no physical danger whatsoever. In actual fact, the "reinforced concrete" hadn't been concrete at all — but mocked-up foam. And the mechanism hadn't jammed. It was perfectly OK. But, of course, I hadn't known that at the time. And neither do the Special Forces candidates who undergo such treatment in selection. From where I was standing (or lying) on the puddle-strewn, diesel-stained floor of some derelict depot in the middle of what could have been anywhere (I'd been taken there hooded to add to my sense of disorientation), it had all been horribly real. Despite what Dave had said about not killing me, when a ten-ton weight is hovering so close you can smell it, and is making it just that little bit more difficult to breathe, it's hard to "unbelieve" that next on the menu is death. Pretty damn close to impossible, in fact. Your brain's so busy running its fear program, it completely overrides its "lie detection" module.

Unbelief is the friction that keeps persuasion in check. Without it, there'd be no limits.

WHEN THE END OF THE WORLD IS *NOT* THE END OF THE WORLD

One of the funny things about unbelief is that sometimes the brain can unbelieve *itself*. Sometimes, when we're really not sure that we like something, or are unhappy with a particular outcome, we convince ourselves it's actually not so bad. And when that starts to happen, next station down is the Mirror Man.

In 1956, the Stanford social psychologist Leon Festinger posed a question that all of us have probably asked ourselves at one time or another: What happens to members of religious cults that prophesy the end of the world . . . and it doesn't happen? Do they all go back to their day jobs and "put it down to experience"? Or what, exactly?

To find out, Festinger infiltrated a UFO doomsday cult — led by Chicago housewife Marion Keech — whose preternatural tip-off that the world, on the morning of December 21, was to be destroyed by alien floodwaters, turned out to be misguided. Not to be out-prophesied, Festinger made a prediction of his own: contrary to what common sense might dictate under such circumstances, the group's proselytizing would — far from petering out after the prophecy malfunction — in contrast actually *increase*. The contradiction of the world ending on the one hand, and life carrying on as normal on the other, would, Festinger proposed, nag the brain into manufacturing a renewed, even stronger, commitment to the cause — so as to reduce the tension between subjective and objective reality, and restore psychological harmony.

This, it transpired, is precisely what happened. Exactly as Festinger foretold, Keech's tip-off turned out *not* to have been misguided after all. Far from it. Rather, her followers came out fighting — with renewed vigor and tighter knit than ever. The marauding aliens, as a gesture to the "true believers," had, it materialized, had second thoughts. The world was to be granted a stay of execution and the entire population spared. It was either that, as Festinger pointed out, or face the unspeakable alternative. That there never *had* been a custom-built flying saucer. That the master plan to spirit them all off into the cosmos had never existed in the first place. And that the jobs, spouses, and houses had all been abandoned in vain.

Festinger's exposé of Keech's divinations precipitated an avalanche of research into the dynamics of *cognitive dissonance*. The flagship study, conducted by Festinger himself in 1959, did much to get things moving. The study consisted of three key ingredients: the obligatory cohort of students, a series of meaningless and mind-numbingly tedious tasks, and a downright whopper of a lie: the students had to perform the tasks and then rope in subsequent "participants" (in reality, associates of the researchers) by claiming that they were actually *interesting*.

The students were split into two groups. Members of one group were paid $1 for their duplicity, the others $20. What effect, Festinger wondered, would the difference in recompense have on the students' *real* ratings of the task?

The answer, it turned out, was a huge one. Sure enough, just as dissonance theory predicted (and totally counter to the laws of common sense) those students who were given just $1 to mislead their fellow participants exhibited *fewer* misgivings about the task than their better-paid counterparts.

Incredible!

And the reason?

According to Festinger, it was simple. The $1 group experienced greater dissonance than the $20 group — $1 as opposed to $20, providing insufficient justification for their attitude-discrepant behavior (telling a person that the tasks were really interesting when in actual fact they were boring). The students, in the absence of any other justification for their behavior, were forced to internalize the attitude they were induced to express — and came, in so doing, genuinely to believe that the tasks they had performed were enjoyable.

On the other hand, however, those in the $20 group had reason to believe there was *external* justification for their behavior — they were in it for the money. No problem there with job satisfaction.

WHY WE LOVE THE THINGS WE HATE
(ESPECIALLY IF WE CAN'T GET A REFUND)

The perils of cognitive dissonance should feature uppermost in the mind of any prospective persuader. Especially in situations where there's a lot at stake and the person whom one is persuading has much to lose. Festinger's study — these days considered a classic — provided, for the very first time, concrete evidence of something that we now take for granted: powerful gravitational forces deep within our brains keep the orbits of both belief and behavior in close psychological alignment. But sometimes, quite evidently, the gravity is too severe — the alignment, in some cases, so slavishly snug that reason disappears into a neural black hole. In advertising, for example, studies have shown that it's not just smokers' arteries that harden as a result of their habit. So, too — in the wake of public health campaigns — do their attitudes.

Consider the dilemma faced by a smoker on exposure to an antismoking ad. The statements "I smoke" and "Smoking kills" are never, for obvious reasons, going to hit it off naturally. They are never going to blossom into the greatest of cognitive room buddies. So either one of them leaves and finds a dorm elsewhere, or they learn to muddle through (the smoker typically focusing on the perceived benefits of their habit — e.g., "It helps me relax"; "All of my friends smoke" — while at the same time downplaying the risks. "Not all experts agree"; "It only affects older people").

The same is true of religious conviction. The cognitive thrift that typi-fies certain believers (as well, of course, as certain nonbelievers) stems from the huge psychological investment loans that such beliefs soak up — often over a period of many years, and often underwritten by those old estab-lished clearing houses of self-identity (moral outlook, social networks, and political affiliation). Could *you* sell up and start again from scratch?

There are other, more mundane examples. Consider what happens when you buy something in a store, have second thoughts about it when you get home — and then, on taking it back, discover that the store em-ploys a "no returns" policy. If you're like most people, what usually happens is this: you magically come round to actually quite *liking* whatever it is that you bought. Hey, you think, as you scrunch up the receipt and toss it into the bin, it's not so bad after all, I guess.

But there's no magic at work here — rather, the hand of cognitive dis-sonance. Two incontrovertible and antithetical cognitions — "I have spent X amount on this particular purchase" on the one hand, and "I don't like it and can't change it" on the other — are forced to cohabit the same bit of brain space until one of two things happens. They either get their act to-gether and sort out their differences. Or one of them packs its bags. Nine times out of ten, they learn to get along.

THE NEUROLOGY OF INFLUENCE

The effects of cognitive dissonance demonstrate quite clearly how the prop-ositional aspects of belief are intricately interwoven with emotion. But a re-cent experiment conducted by Sam Harris and his colleagues at the Univer-sity of California in Los Angeles goes one better — and shows how belief, emotion, and influence are intricately interwoven in the brain.

Using a special goggle display unit which participants wore over their eyes, Harris flashed statements from seven different subject categories, which they rated for authenticity. Each of the seven categories (mathemat-ical, geographical, autobiographical, religious, ethical, semantic, and fac-tual) contained three kinds of statement: statements that were true, state-ments that were false, and statements that were neither — those, in other words, that couldn't be verified either way. (For instance, an example of a *true/mathematical* statement might be $(2 + 6) + 8 = 16$; an example of a *false/ethical* statement might be "Children should have no rights until they vote"; and an example of a *religious/unverifiable* statement might be "Jesus spoke 2,467 words in the New Testament.")

As participants evaluated the statements, Harris turned detective, snooping around inside their brains using fMRI. Which anatomical regions, he wondered, corresponded to each of the different appraisals — to the belief, disbelief, and uncertainty that the three kinds of statement elicited?

The results were intriguing. To begin with, reaction-time data revealed that statements were accepted as true faster than they were rejected as false — providing further support for Spinoza's original conjecture that first we believe, and then "unbelieve."

But there was more. Belief, Harris found, was accompanied by an increase in activity in the ventromedial prefrontal cortex (see Figure 8.1a) — the part of the brain usually associated with integrating fact and feeling, and with modulating behavior in response to changing reward contingencies (weighing up pros and cons, in other words). Disbelief, in turn, activated the anterior insula (see Figure 8.1b, overleaf) — the region often implicated in the coding of aversive reactions such as pain and disgust, and in assessing the pleasantness of different tastes and odors. Uncertainty, as predicted, nudged the anterior cingulate cortex — a kind of neurological warning light that flashes on and off when something perplexingly novel suddenly appears on the radar (see Figures 8.1c and 8.1d, overleaf).

Figure 8.1a — Activation of the ventromedial prefrontal cortex regions for judgments of truth (belief), across the seven statement categories: mathematical, geographical, autobiographical, religious, ethical, semantic, and factual. Sections in white indicate areas of increased brain activity.

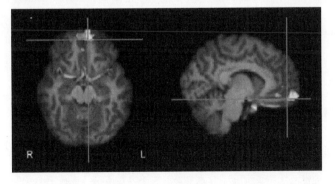

Figure 8.1b — Axial image (left) shows increased activity in the inferior frontal gyrus (primarily the left), the right middle frontal gyrus, and the interior insula (bilateral) for judgments of falsity (disbelief) across the seven statement categories. Sagittal image (right) shows increased activity in the superior parietal lobe, the cingulate cortex, and the superior frontal gyrus.

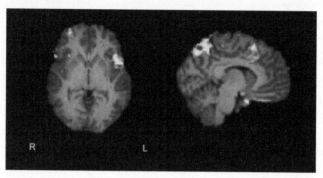

Figure 8.1c (top) and 8.1d (bottom) — Activation of the anterior cingulate and superior frontal gyrus during judgments of uncertainty. Figure 8.1c shows the contrast with judgments of belief. Figure 8.1d shows the contrast with disbelief.

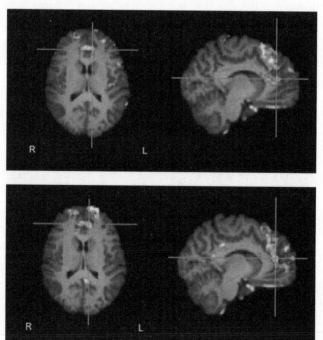

Could these be the regions that define the limits of influence? Might beliefs that fire the emotions, that crucially give rise to a heightened level of activity in the ventromedial prefrontal cortex, be particularly hard to shift? While those that antagonize the prickly anterior insula are particularly hard to acquire? The hypothesis certainly seems plausible — though as Mark Cohen, one of Harris's coauthors, pointed out when I put it to him, it's one thing looking at neural correlates in the lab and quite another, say, in the boardroom. Where people are involved. And the propositional markers of "true," "false," and "don't know" are considerably less sanitized.

"Persuasion is a social thing," he says. "And social interaction introduces a number of other brain circuits not specifically catered for in our study. . . . But what we *can* say is that belief, disbelief, and uncertainty do seem to be associated with discrete neural signatures of acceptance, rejection, and indecision."

For SPICE, too, there are implications. A style of persuasion that engages, simultaneously, all three of the brain's influence hotspots (incongruity — the anterior cingulate cortex; simplicity, perceived self-interest, confidence, and empathy — the ventromedial prefrontal cortex; with the combination of all five elements *disengaging*, rather than engaging, the redundant anterior insula) is undoubtedly going to be powerful. Under some circumstances (one thinks, for instance, of neonatal persuasion: crying pitch activating the anterior cingulate cortex, and *kindchenschema* networks in the prefrontal cortex) irresistible even.

DRIVEN TO DISTRACTION

In blazing 90-degree heat, on the banks of the Swan River, over a beer, I discuss Harris's findings with Colin MacLeod, professor of clinical psychology at the University of Western Australia, in Perth. MacLeod is an expert in anxiety disorders, and knows only too well how belief and emotion can sink their teeth into each other. He is about to introduce me to Tania — a twenty-seven-year-old manicurist with a seatbelt phobia, who works at a local beauty salon. Or rather did, until she was forced into selling her car.

"A lot of the time we worry about the worry," explains MacLeod. "We conflate the thing we're worrying about with the worry that the thing we're worrying about gives rise to. This 'second order' worry then takes over and things get confused. The second order worry gradually becomes the focal point of the problem — the first order worry, if you get my meaning.

"So ironically, what we're going to be doing with Tania is getting her to focus her anxiety on the seatbelt, because in doing so what we're actually doing, unbeknown to her, is taking her mind off the *real* source of her anxiety — the worry about the worry — and transplanting it onto a 'ghost' anxiety: the original hub that now lies emotionally dormant. Basically it's distraction in disguise. What Tania will be unconsciously 'concentrating away' will not, in fact, be the actual phobia *itself* — but a satellite anxiety associated with the phobia's onset."

Catch the brain with its pants down, and anything is possible.

As Tania arrives and we set off for the parking lot, MacLeod begins chatting to her. Putting her at ease.

"What I need to do first," he explains matter-of-factly, "is to see the symptoms firsthand — for myself — in order to know what I can do about them. Is that OK?"

Tania nods.

"Good," says MacLeod. "So let's do it in stages. First, tell me how you feel right now, as we approach the parking lot. Concentrate hard on those feelings of anxiety and try to express them to me."

Tania goes quiet for a few seconds as she tries to sum up how she feels. Then she says, "Well, actually, I seem to be OK at the moment."

"OK," says MacLeod. "That's fine. We'll try again in a minute."

Sure enough, as we reach her car, MacLeod asks Tania the same question. Focus on that anxiety, he urges her, and tell me how it feels. Once again, Tania draws a blank. Amazingly, she draws another blank a few seconds later as she is getting into the car. And another as she puts on her seatbelt. Driving around the parking lot doesn't seem to cause her any problems. And neither does the traffic on the freeway. Suddenly it appears that the appointment has been unnecessary. The symptoms a false alarm. And that the phobia — if, indeed, that's what it is — has never really existed. Except, of course, that it has. And it's recently cost her her job.

Back on the river, over another beer, I suggest to MacLeod that what he's just used is SPICE. He doesn't disagree — though the technical term, he points out, is actually *paradoxical intention*: the eradication of a particular symptom by making the symptom *itself* the sole focus of attention. This gets me thinking. Before my meeting with MacLeod, it hadn't really occurred to me that therapy was a form of persuasion. I guess because it's "medical." And you have to make an appointment. But MacLeod himself is under no such illusions.

"Therapy is *absolutely* about persuasion," he says. "Fundamentally, it's about changing people's belief systems. And the bottom line is that therapists are professional persuaders. What CBT* — my own brand of therapy — does is it enacts a paradigm shift in people's heads. It doesn't come up with the *solution* to the problem but rather a different way of *thinking* about it. It isn't so much about providing a key. But what it does do is persuade the client to think about changing the lock."

BE HAPPY, DON'T WORRY

Over the past few years, MacLeod has been at the forefront of a brand new form of therapy called cognitive bias modification (CBM) which, if it works (and the early signs are good), could completely redefine the limits of persuasion. As a postdoc back in the early eighties, MacLeod was among the first wave of researchers to bring the methods of cognitive psychology to the *clinical* table — specifically, to the area of anxiety disorder. What were anxious people *thinking*? MacLeod wanted to know. And how did it differ from what the rest of us were thinking? What he came up with was profound. Just as, say, a Manchester United fan will zone in on the words "Manchester United" on a page of otherwise irrelevant text, so the attention of anxious individuals is drawn, inexorably, to threatening things around them. Unlike the rest of us, who are able to screen them out, anxious people can't. They are, to use the technical term, "threat vigilant."

MacLeod has demonstrated this using a paradigm known as the dot probe task. Participants are divided into two groups — anxious and nonanxious — and stare at a fixation cross at the center of a computer screen. Two words, one of which is neutral and one of which is threatening, are then presented randomly on either side of the screen (left or right) for around 500 milliseconds, before a probe (usually a dot) appears in one of these former locations. Participants must then indicate the location of this dot (left or right) as quickly as possible via a key press — before repeating the process over a series of subsequent trials.

When, at the end of the procedure, reaction times are averaged and the performance of the anxious and nonanxious groups is compared, a telltale difference materializes. Anxious individuals, it emerges, are faster at locating the probe when it appears in a position formerly occupied by a threat-

---

*Cognitive behavioral therapy.

ening word than they are when it appears in a neutral position — a disparity not found among the nonanxious. Anxious individuals, in other words,
have a *cognitive bias* toward threat.

Recently, MacLeod has been thinking in a different way about the dot
probe paradigm. At the outset, as we've just seen, the procedure was instrumental in uncovering precisely what it was that was *driving* anxiety — at
least, that is, on a cognitive level. But might it also have the power to *reduce*
it? To "concentrate" the threat bias away? MacLeod believes that it does. Not
only that, he's also got the evidence to prove it.

In 2002, he and his coworkers modified the paradigm so that the probe
was no longer random. That is, rather than pop up with equal frequency in
the threat and neutral locations, it appeared 100 percent of the time in either one or the other location: in the case of a threat-related word, the *attend threat* (AT) condition; in the case of a neutral word, the *attend neutral*
(AN) condition. MacLeod then selected a bunch of volunteers with midrange anxiety scores (as measured by a standardized anxiety questionnaire)
and divided them into two groups. One group got six hundred blasts of AT,
the other six hundred blasts of AN.

Could the dot probe be transformed from an experimental paradigm
into a *training* paradigm? MacLeod wondered. Was it possible, through
repeated attentional direction to one or other location, to actually *induce*
biases?

The answer, it turned out, was yes. When, at the conclusion of the "training," those in the AT group were presented with a normal dot probe task,
they showed — guess what? — increased vigilance for threat-related words.
In contrast, those in the AN group showed increased vigilance for neutral
words. Not only that, but when the volunteers were given a subsequent ana-
gram-solving test specifically designed to make them feel anxious (most of
the anagrams might just as well have been in Swahili, and some were actually impossible), those in the AN group exhibited fewer signs of stress than
those in the AT group.

And the story doesn't end there. While MacLeod has been working on
attentional training, Andrew Mathews and Bundy Mackintosh at the MRC
Cognition and Brain Sciences Unit in Cambridge have been thinking along
similar lines — and developing a technique to modify how we *interpret* situations. From his years as a clinical psychologist at St. George's Hospital in
London, Mathews has realized that just as the *attention* of anxious individ-

uals is drawn to threatening stimuli in the environment, so, too, are their *thought processes*. Whereas the rest of us might look on the bright side, anxious individuals, as a rule of thumb, tend to do the opposite — they interpret things in a negative, more hostile fashion. To illustrate, Mathews provides an example. Once, as a teenager, a colleague of his developed a large pimple on his face the day he was going out with a new date. Infuriated by the less than sympathetic reaction of his younger brother, he stormed out of the house and sat on a bench on top of a nearby mound overlooking the town. Five minutes later, a tourist came and sat next to him.

"Nice spot," he said. . . .

Mathews's approach is the same as MacLeod's — except that instead of training attention, he trains cognition. In a typical experiment, volunteers are presented with a bunch of scenarios that must be resolved either positively or negatively by completing a word fragment at the end. This comprises the training phase.

For example: "Your partner asks you to go to an anniversary dinner that his company is holding. You have not met any of his work colleagues before. Getting ready to go, you think that the new people you will meet will find you . . ."

In the condition designed to induce a *negative* interpretation bias, the word fragment to be completed would be "bo--ng" (boring) — and you'd work through one hundred such examples. In the *positive* bias condition, it would be "fri---d-y" (friendly) — and you'd get one hundred of *those*.

Subsequently, in the *test* phase, volunteers are then presented with another batch of scenarios — similar to the first — except that this time the conclusions remain ambiguous, and are accompanied by a number of possible outcomes which must be rated for goodness of fit.

Exactly as in MacLeod's attentional training procedure, Mathews finds that those volunteers who are trained to interpret things negatively endorse the outcomes most consonant with that bias — the same being true of those in the positive condition. Not only that, but on exposure to subsequent stressors (e.g., video clips of injuries and accidents), those who've undergone the positive training exhibit less anxiety than those who've been given the negative.

"Don't worry, be happy," sang Bobby McFerrin. But it should be the other way round.

PERSUASION PATHWAYS

MacLeod and Mathews are optimistic about the future. (Well, they would be, wouldn't they?) And with good reason. If, as Sam Harris showed us, belief is a brain state, then by changing brain states we should, in theory, be able to change beliefs. Not just in theory, but in fact. And not just some beliefs, but all of them. Religious, political, you name it.

In 2004, a couple of years after his initial modification of the dot probe paradigm, MacLeod used exactly the same procedure with sufferers of social phobia. Over a two-week period, patients received a daily dose of 384 training trials explicitly diverting their attention away from threat-related words. Result? A significant reduction in symptoms.

A year later, in 2005, Matt Field and Brian Eastwood at the University of Liverpool adapted CBM for use with heavy drinkers (neutral images vs. alcohol-related images). On an ingenious follow-up measure of alcohol dependency, they found that those in the "attend neutral" condition sampled less beer in a "taste test" than those in the "attend alcohol" condition.

Even more spectacular are studies that have been done on stroke victims. Edward Taub, at the University of Alabama, has set up the Taub Therapy Clinic where it's not unusual to see patients weighed down by mitts and slings on their *good* limbs. The reason for this is not at all obvious — unless you've experienced the kind of "Aha!" moment that Taub has.

Taub has discovered that the brains of stroke victims go into a state of "cortical shock" after their initial seizure, during which time any attempt to move affected limbs is met with failure. Over a period of several months, the result of such failure entails what Taub refers to as "learned nonuse" (a variant on the learned helplessness we encountered in Chapter 5) — whereby the neural motor map of the stricken body part (in accordance with the brain's immutable "use it or lose it" principle) begins to atrophy. But force an individual to work the affected area, to persevere in the face of repeated failure (hence the mitts and slings — or the accoutrements of what Taub calls "constraint-induced" movement therapy), and remarkable strides, quite literally, can be made. The brain can be taught to rewire itself: to pack off new generations of dendrites, across uncharted neural landscapes, in the footsteps of the old. And if you can "persuade away" paralysis, then who knows where such influence ends?

Elaine Fox, professor of psychology at the University of Essex, is taking things one stage further than MacLeod and Mathews and, using fMRI, is looking at the effects of CBM deep within the brain. Her research program (in collaboration with Naz Derakshan at the University of London) is currently in its infancy, but one of the areas she'll be keeping a close eye on is the ventromedial prefrontal cortex — which codes, as Sam Harris demonstrated, for the formation of beliefs. In particular, she'll be watching out for changes in the attentional control networks between the prefrontal cortex and the amygdala — and if anything comes up, may well be on her way to solving persuasion's version of the "hard problem": by isolating, for the very first time, a discrete "persuasion pathway" in the brain.

"It's not persuasion in the strictest sense of the word," says Fox, "because in CBM the individual is a willing participant in the belief change process, and the contingencies of the procedure are subliminal. But as an indicator of what changes in the brain when we change our minds, it's certainly a start."

It certainly is. Whether it's an open or a fixed mind-set; whether the mothership's going to come and get you, or whether you pick up your bed and walk — the code for each of these belief systems is encrypted in the brain in ancient mathematical lightning storms that fork across its surface in milliseconds. Divert the course of these electrochemical data shoals, or modify their intensity, and you'll navigate belief along shadowy meridians of influence toward change.

You will, in other words, *persuade.*

Back in Sydney, in the University of Macquarie's Centre for Cognitive Science, I probe a little deeper about the Mirror Man.

"What if he stands over a puddle and strikes a match?" I ask Max Coltheart, founder of the Belief Formation Programme. "He'll then have to explain how his alter ego has accomplished the same feat under water."

Coltheart shrugs. He's heard it all before.

"Well, he's managed to explain how his shaving companion has been following him around the bathroom all this time," he says. "He's even caught him in bed with his wife! So he'll sure as hell come up with something. The mistake people make is to think that the answer is somehow rooted in logic. It isn't. He's said himself that he knows what he says sounds crazy. The problem lies in how his brain makes sense of the world. How it orders the sensory data and tries to construct an internal, coherent narrative."

And therein lies the secret.

"Cases are won and lost not just on the strength of facts," the lawyer Michael Mansfield told us in Chapter 4, "but on impressions. A lot is achieved through the power of suggestion. . . . It's not just about presenting the evidence. It's about *how* you present it."

The jury's still out in Sydney.

POSTSCRIPT — PERFECT IMPERFECTION

One of the questions I'm often asked about split-second persuasion is whether anyone can do it. Do we all have the ability to wheel and deal at the flipping points of life or is it just the prerogative of a privileged few — persuasion geniuses with a special kind of know-how?

My answer is always the same.

The issue is one of degree. Most, if not all, of us have a dial-up connection to some imperishable realm of pure Platonic perfection. And most, if not all, of us will have occasionally got through to it by accident. When was the last time you said exactly the *right* thing at exactly the *right* time — only to discover so later? You might not have known it at the time — but, hey, that's often what makes it right!

Equally impressive, though dialed a little more frequently, is the direct line to Platonic *imperfection*. When was the last time you said exactly the *wrong* thing at the right time?

Easier to remember, no?

And I bet you found out quicker.

In the lead-up to Christmas, the Royal Mail receives in excess of 750,000 letters from children all over the British Isles addressed to Santa. There are strict regulations for dealing with this kind of correspondence — and those petitions that don't accidentally slip into the recycling unit are carefully filed away. But a couple of years ago one letter in particular succeeded in catching the eye of a female employee in one of the regional sorting offices. It was from a little boy who'd been saving up all year for a PlayStation — but who was only halfway there. His mother was ill and his father had just been laid off. So the family, as you can imagine, was on a pretty tight budget. Could Santa come to the rescue (to the tune of £200)?

The woman in the sorting office who opened the letter passed it around among her colleagues. They were all extremely moved. In fact, so touched were they by the young lad's hard work and enterprise — he'd been washing cars and doing a couple of paper routes — that they decided to hold a collection for him. Everyone contributed generously and by the time they'd finished a big fat envelope with £120 in it sat on the woman's desk. This she dispatched to the diligent little fella — with an accompanying note from "Santa" wishing both him and his family all the very best for the New Year.

And that, pretty much, was that. Nothing more came of it. Until, that is, several weeks later in mid-January when a letter addressed to Santa turned up at the same sorting office. The woman who'd dealt with the first letter also handled the second. This is what she read:

> Dear Santa,
>   Many thanks for your Christmas gift of £200 for my son. It was very generous of you. Unfortunately, however, he's still not managed to get hold of the PlayStation he wanted because when he opened your letter he found that there was only £120 inside it. Those thieving bastards at the Post Office must have creamed off £80 for themselves. I suppose it just goes to show you can't trust anyone these days.

Ouch!

Some people, it would seem, are simply not content with getting the wrong end of the stick. They insist, instead, on getting the wrong *stick*. We've all been there, haven't we? If experience teaches us anything, it's this: behind the façade of assiduous, fumbling accomplishment there shimmers a realm of despicably effortless incompetence. An imperishable array of faux pas, cock-ups, and howlers that clunks into mortal existence at the whim of the cognitively challenged.

One evening, outside Cambridge railway station, I found myself muttering in a long, disgruntled queue that had gradually been taking shape by the taxi stand like a slow-moving tropical cyclone. Suddenly, out of nowhere, a loudmouth, teenage pisshead sauntered nonchalantly up to the front. With laudable restraint, the guy standing next to me politely called him aside. And invited him, as cordially as the circumstances permitted, to fuck off back down the line. But the impertinent interloper wasn't having any of it.

"I've just had a call to say that my girlfriend's been rushed to hospital," he slurred. "And that she's going straight into surgery. What's your excuse?"

"I'm the surgeon," came the reply.

Makes you wonder, doesn't it?

If we can get it so staggeringly wrong — then what is there to stop us from getting it so staggeringly *right*?

Multidimensional Iowa Suggestibility Scale (MISS) (Short Version)

Please indicate to what extent the following statements apply to you. Use the following scale to record your answers, then add up your total at the end:

1 — Not at all or very slightly
2 — A little
3 — Somewhat
4 — Quite a bit
5 — A lot

1. I am easily influenced by other people's opinions.
2. I can be influenced by a good commercial.
3. When someone coughs or sneezes, I usually feel the urge to do the same.
4. Imagining a refreshing drink can make me thirsty.
5. A good salesperson can really make me want their product.
6. I get a lot of good practical advice from magazines or TV.
7. If a product is nicely displayed, I usually want to buy it.
8. When I see someone shiver, I often feel a chill myself.
9. I get my style from certain celebrities.
10. When people tell me how they feel, I often notice that I feel the same way.
11. When making a decision, I often follow other people's advice.
12. Reading descriptions of tasty dishes can make my mouth water.
13. I get many good ideas from others.
14. I frequently change my opinion after talking with others.
15. After I see a commercial for lotion, sometimes my skin feels dry.
16. I discovered many of my favorite things through my friends.
17. I follow current fashion trends.
18. Thinking about something scary can make my heart pound.
19. I have picked up many habits from my friends.
20. If I am told I don't look well, I start feeling ill.
21. It is important for me to fit in.

SCORING: **20–40** You're as tough as nails. No definitely means no. **40–60** You're no pushover. You know your own mind and are not easily swayed. **60–75** You're open to offers and will often "give it a go." **75+** Can I interest you in a little deal I've got going at the moment?

QUESTIONNAIRE SUBSCALES: **Physiological Suggestibility**—Items 8, 10, 15, 20, 3. **Consumer Suggestibility**—Items 2, 9, 5, 6, 7. **Peer Conformity**—Items 19, 17, 21, 16. **Physiological Reactivity**—Items 18, 4, 12. **Persuadability**—Items 14, 1, 13, 11.

(MISS. Copyright © 2004 by R. I. Kotov, S. B. Bellman, and D. B. Watson.)

APPENDIX 1
Key Stimuli and Stereotyping:
Socioeconomic Status

APPENDIX 2
Asch's Supplementary Traits

NOTES

ACKNOWLEDGMENTS

ILLUSTRATION CREDITS

INDEX

Key Stimuli and Stereotyping: Socioeconomic Status

The following two character sketches differ only in the last item. In Sketch A, Mr. Jones lives in a large house with a pool; in Sketch B, he lives in a high-rise apartment block:

SKETCH A	SKETCH B
1. Mr. Jones is forty-three years old.	Mr. Jones is forty-three years old.
2. He is married and has two children.	He is married and has two children.
3. His hobbies include horse-racing and working out.	His hobbies include horseracing and working out.
4. He usually takes his holidays in Florida.	He usually takes his holidays in Florida.
5. He lives in a large house in the country.	He lives in a high-rise apartment block.

Give one group of friends Sketch A and another group Sketch B, and ask them to form an impression in their mind of the kind of person Mr. Jones is likely to be.

When they've had a moment or two to clarify their thoughts, give them the following impression-formation task and note the pattern of responses for each group.* You should see quite a difference!

---

*Make sure that members of the two groups don't confer when answering the questions!

Out of each pair of statements below, indicate which one you think is more likely to apply to Mr. Jones.

STATEMENT A	STATEMENT B
1. Mainly an optimist	Mainly a pessimist
2. Takes his work lightly	Conscientious in his work
3. Spends time with his children	Leaves his children to their own devices
4. Good with money	Reckless with money
5. Rarely does housework	Often does housework
6. Lives mainly in the present	Plans for the future
7. Attentive to his wife	Tends to take his wife for granted
8. Fairly fond of gambling	Opposed to gambling
9. Self-reliant	Dependent on others
10. Somewhat untidy	Meticulous in his habits
11. Largely self-centered	Great concern for others
12. Active church member	Not bothered about religion
13. Loud and boisterous	Quiet and reserved
14. Shares his wife's interests	Husband and wife do their own thing
15. Democrat	Republican
16. Slow and deliberate	Quick and impulsive
17. Somewhat ambitious	Has few ambitions
18. Rather patriotic	Not very patriotic
19. On friendly terms with neighbors	Tends to keep himself to himself
20. Scrupulously honest	Not averse to turning a blind eye

Of course, where you live constitutes only one example of the kind of information that influences social perception. By experimenting with the above format — varying the items in both the character sketches and on the impression-formation task — it's possible to uncover quite a few others.

Asch's Supplementary Traits

Having read their "warm/cold" character descriptions, Asch's participants then selected which adjective from the following eighteen trait pairs was most in accordance with the view they'd formed of the individual:

1. Generous — Ungenerous
2. Shrewd — Wise
3. Unhappy — Happy
4. Irritable — Good-natured
5. Humorous — Humorless
6. Sociable — Unsociable
7. Popular — Unpopular
8. Reliable — Unreliable
9. Important — Insignificant
10. Ruthless — Humane
11. Good-looking — Unattractive
12. Persistent — Unstable
13. Frivolous — Serious
14. Restrained — Talkative
15. Self-centered — Altruistic
16. Imaginative — Hardheaded
17. Strong — Weak
18. Dishonest — Honest

Below is the frequency (in terms of percentages) with which each item in the checklist was selected. (*Note:* Only the results for the *positive* term in each pair are given. To determine the percentage of the *negative* item, subtract the given figure from 100.)

	"WARM" ($N = 90$)	"COLD" ($N = 76$)
Generous	91	8
Wise	65	25
Happy	90	34
Good-natured	94	17
Humorous	77	13
Sociable	91	38
Popular	84	28
Reliable	94	99
Important	88	99
Humane	86	31
Good-looking	77	69
Persistent	100	97
Serious	100	99
Restrained	77	89
Altruistic	69	18
Imaginative	51	19
Strong	98	95
Honest	98	94

Notes

page 1. THE PERSUASION INSTINCT

15 *McComb and her coworkers compared cat owners' responses . . .* : McComb, Karen, Anna M. Taylor, Christian Wilson, and Benjamin D. Charlton. "The Cry Embedded within the Purr." *Current Biology* 19(13) (2009): R507–508.

17 *In certain species of frog it is sound . . .* : "Louisiana's State Amphibian, the Green Treefrog." www.americaswetlandresources.com/wildlife_ecology/plants_animals_ecology/animals/amphibians/GreenTreeFrogs.html (accessed June 5, 2008).

18 *As a weapon of persuasion . . .* : For more on mimicry and deception see Peter Forbes, *Dazzled and Deceived: Mimicry and Camouflage* (London: Yale University Press, 2009).

19 *The discomycete fungus . . .* : Ngugi, Henry K., and Harald Scherm. "Pollen Mimicry during Infection of Blueberry Flowers by Conidia of *Monilinia vaccinii-corymbosi.*" *Physiological and Molecular Plant Pathology* 64(3) (2004): 113–123.

19 *Conversely, Owl Butterflies . . .* : For an alternative interpretation, see Martin Stevens, Chloe J. Hardman, and Claire L. Stubbins, "Conspicuousness, Not Eye Mimicry, Makes 'Eyespots' Effective Antipredator Signals," *Behavioral Ecology* 19(3) (2008): 525–531.

20 *The Golden Orb Weaver . . .* : Théry, Marc, and Jérôme Casas. "The Multiple Disguises of Spiders: Web Colour and Decorations, Body Colour and Movement." *Philosophical Transactions of the Royal Society* B 364 (2009): 471–480.

21 *Studies have shown that female fireflies . . .* : Lloyd, James E. "Aggressive Mimicry in Photuris: Firefly Femmes Fatales." *Science* 149 (1965): 653–654; Lloyd, James E. "Aggressive Mimicry in Photuris Fireflies: Signal Repertoires by Femmes Fatales." *Science* 187 (1975): 452–453.

22 *In fact, research conducted in 2001 . . .* : McCleneghan, J. Sean. "Selling Sex to College Females: Their Attitudes about *Cosmopolitan* and *Glamour* Magazines." *Social Science Journals* 40(2) (2003): 317–325.

23 *Herring Gull chicks instinctively respond . . .* : Tinbergen, Nikolaas, and Albert C. Perdeck. "On the Stimulus Situation Releasing the Begging Response in the Newly Hatched Herring Gull Chick (*Larus argentatus* Pont.)." *Behaviour* 3 (1950): 1–38.

27 *Recent work on crayfish* . . . : Issa, Fadi A., and Donald A. Edwards. "Ritualized Submission and the Reduction of Aggression in an Invertebrate." *Current Biology* 16 (2006): 2217–2221. For a practical guide to understanding and interpreting nonverbal communication cues (including those of sales reps!) see Gordon R. Wainwright, *Body Language* (London: Hodder Education, 2003).

30 *In the summer of 1941* . . . : www.anecdotage.com/browse.php?category=people &who=Churchill (accessed April 2, 2008).

2. FETAL ATTRACTION

32 *A Houston lady just told me* . . . : "Cry Baby." www.snopes.com/crime/warnings/ crybaby.asp (accessed March 9, 2008).

33 *Studies have shown that infants* . . . : McCall, Robert B., and Cynthia Bellows Kennedy. "Attention of 4 Month Infants to Discrepancy and Babyishness." *Journal of Experimental Child Psychology* 29(2) (1980): 189–201.

34 *More surprising still.* . . . : Sackett, Gene P. "Monkeys Reared in Isolation with Pictures as Visual Input: Evidence for an Innate Releasing Mechanism." *Science* 154 (1966): 1468–1473.

34 *Research conducted by neuroscientist Morten Kringelbach* . . . : Kringelbach, Morten L., Annukka Lehtonen, Sarah Squire, Allison G. Harvey, Michelle G. Craske, Ian E. Holliday, Alexander L. Green, Tipu Z. Aziz, Peter C. Hansen, Piers L. Cornelissen, and Alan Stein. "A Specific and Rapid Neural Signature for Parental Instinct." *PloS ONE* 3(2) (2008): e1664.

35 *In 1998, the Pentagon commissioned Pam Dalton* . . . : Pain, Stephanie. "Stench Warfare." New Scientist Science Blog (July 2001). www.scienceblog.com/community/ older/2001/C/200113657.html (accessed November 18, 2005).

36 *Compared to the racket of a three-pronged garden pitchfork* . . . : For a detailed acoustical analysis of the properties of an aversive sound, see Sukhbinder Kumar, Helen M. Forster, Peter Bailey, and Timothy D. Griffiths, "Mapping Unpleasantness of Sounds to Their Auditory Representation," *Journal of the Acoustical Society of America* 124 (6) (2008): 3810–3817.

36 *British inventor Howard Stapleton* . . . : Jha, Alok. "Electronic Teenager Repellant and Scraping Fingernails, the Sounds of Ig Nobel Success." *Guardian,* October 6, 2006. www.guardian.co.uk/uk/2006/oct/06/science.highereducation (accessed October 28, 2006).

37 *The normal range of adult human hearing* . . . : For a detailed analysis of infant crying see Joseph Soltis, "The Signal Functions of Early Infant Crying," *Behavioral and Brain Sciences* 27 (2004): 443–490; and Debra M. Zeifman, "An Ethological Analysis of Human Infant Crying: Answering Tinbergen's Four Questions." *Developmental Psychobiology* 39 (2001): 265–285.

38 *In 2007, Kerstin Sander* . . . : Sander, Kerstin, Yvonne Frome, and Henning Scheich. "FMRI Activations of Amygdala, Cingulate Cortex, and Auditory Cortex by Infant Laughing and Crying." *Human Brain Mapping* 28 (2007): 1007–1022.

39 *Paul Rozin and his colleagues* . . . : Rozin, Paul, Alexander Rozin, Brian Appel, and Charles Wachtel. "Documenting and Explaining the Common AAB Pattern

in Music and Humor: Establishing and Breaking Expectations." *Emotion* 6(3) (2006): 349–355.

41 *"Music," writes V. S. Ramachandran . . .* : Ramachandran, V. S., and William Hirstein. "The Science of Art: A Neurological Theory of Aesthetic Experience." *Journal of Consciousness Studies* 6 (1999): 15–51.

41 *David Huron, in his book . . .* : This quote taken from Lauren Stewart, "Musical Thrills and Chills," *Trends in Cognitive Sciences* 11 (2007): 5–6.

42 *(Footnote) Such findings may be explained . . .* : For early studies on the halo effect see Solomon E. Asch, "Forming Impressions of Personality," *Journal of Abnormal and Social Psychology* 41 (1946): 258–290; and Edward L. Thorndike, "A Constant Error on Psychological Rating," *Journal of Applied Psychology* 4 (1920): 25–29.

42 *Mark Snyder at the University of Minnesota . . .* : Snyder, Mark, Elizabeth D. Tanke, and Ellen Berscheid. "Social Perception and Interpersonal Behaviour: On the Self-Fulfilling Nature of Social Stereotypes." *Journal of Personality and Social Psychology* 35 (1977): 656–666.

43 *(Footnote) Just in case you were wondering . . .* : Andersen, Susan M., and Sandra L. Bem. "Sex Typing and Androgyny in Dyadic Interaction: Individual Differences in Responsiveness to Physical Attractiveness." *Journal of Personality and Social Psychology* 41 (1981): 74–86.

43 *In 2007, in a study involving lap dancers . . .* : Miller, Geoffrey, Joshua M. Tybur, and Brent D. Jordan. "Ovulatory Cycle Effects on Tip Earnings by Lap Dancers: Economic Evidence for Human Estrus?" *Evolution and Human Behavior* 28 (2007): 375–381.

44 *Figure 2.3 . . .* : Little, Anthony C., and Peter J. B. Hancock. "The Role of Distinctiveness in Judgements of Human Male Attractiveness." *British Journal of Psychology* 93(4) (2002): 451–464.

44 *David Perrett, on the other hand . . .* : Penton Voak, Ian S., David I. Perrett, Duncan L. Castles, Tessei Kobayashi, D. Michael Burt, Lindsey K. Murray, and Reiko Minamisawa. "Menstrual Cycle Alters Face Preference." *Nature* 399 (1999): 741–742.

46 *In 1943, in his classic paper . . .* : Lorenz, Konrad. "Die angeborenen Formen möglicher Erfahrung [The Innate Forms of Potential Experience]." *Zeitschrift fur Tierpsychologie* 5 (1943): 235–409.

46 *Figure 2.5 . . .* : Lorenz, Konrad. "Part and Parcel in Animal and Human Societies: A Methodological Discussion, 1950." In *Studies in Animal and Human Behaviour*, vol. 2. London: Methuen, 1971.

47 *Consider, for example, the series of craniofacial profiles . . .* : Pittenger, John B., and Robert E. Shaw. "Aging Faces as Viscal Elastic Events: Implications for a Theory of Nonrigid Shape Perceptions." *Journal of Experimental Psychology: Human Perception and Performance* 1(4) (1975): 374–382.

47 *Take a look at Figure 2.7 . . .* : Pittenger, John B., Robert E. Shaw, and Leonard S. Mark. "Perceptual Information for the Age Level of Faces as a Higher Order Invariant of Growth." *Journal of Experimental Psychology: Human Perception and Performance* 5(3) (1979): 478–493.

48 *In 2009, Melanie Glocker . . .* : Glocker, Melanie L., Daniel D. Langleben, Kosha Ruparel, James W. Loughead, Jeffrey N. Valdez, Mark D. Griffin, Norbert Sachser,

and Ruben C. Gur. "Baby Schema Modulates the Brain Reward System in Nullip-arous Women." *Proceedings of the National Academy of Sciences* 106(22) (2009): 9115–9119.

48 *Psychologist Richard Wiseman . . .* : Devlin, Hannah. "Want to Keep Your Wallet? Carry a Baby Picture." *Times Online,* July 11, 2009. www.timesonline.co.uk/tol/news/science/article6681923.ece (accessed July 18, 2009).

49 *Another study in America evoked a similar kind of protectiveness . . .* : King, Laura A., Chad M. Burton, Joshua A. Hicks, and Stephen M. Drigotas. "Ghosts, UFOs, and Magic: Positive Affect and the Experiential System." *Journal of Personality and Social Psychology* 92(5) (2007): 905–919.

49 *Sheila Brownlow and Leslie Zebrowitz . . .* : Brownlow, Sheila, and Leslie A. Ze-browitz. "Facial Appearance, Gender, and Credibility in Television Commer-cials." *Journal of Nonverbal Behaviour* 14 (1990): 51–60.

50 *In politics, too . . .* : Brownlow, Sheila. "Seeing Is Believing: Facial Appearance, Cred-ibility, and Attitude Change." *Journal of Nonverbal Behavior* 16 (1992): 101–115.

50 *Back in 2008, a team from the University of Kent . . .* : Gill, Charlotte. "Fresh-Faced Cameron Beats Sunken-Eyed Brown on 'Face You Can Trust' Issue." *Mail On-line,* November 17, 2008. www.dailymail.co.uk/news/article-1086396/Freshfaced CameronbeatssunkeneyedBrownfacetrustissue.html (accessed January 8, 2009).

51 *In relationships, women are more likely to confide . . .* : For more on the pros and cons of baby-faced-ness see Leslie A. Zebrowitz, "The Boons and Banes of a Babyface," in *Reading Faces: Window to the Soul?* chap. 6 (Boulder, CO: Westview Press, 1997).

51 *Take the four photographs of military cadets . . .* : Mazur, Allan, Julie Mazur, and Caroline Keating. "Military Rank Attainment of a West Point Class: Effects of Cadets' Physical Features." *American Journal of Sociology* 90 1 (1984): 125–150. Cadet photographs from The Howitzer, 1950; later career photographs from the U.S. Army Military History Institute and the Center for Air Force History. Images supplied with permission of Professor Allan Mazur, Syracuse University ©.

55 *Statistics like these . . .* : For a detailed discussion of the dynamics of eye con-tact see Michael Argyle, *The Psychology of Interpersonal Behaviour,* 4th ed. (Har-mondsworth: Penguin, 1983); and Albert Mehrabian, *Silent Messages: Implicit Communication of Emotions and Attitudes* (Belmont, CA: Wadsworth, 1971).

56 *In 2007, a team from the University of Geneva . . .* : Brosch, Tobias, David Sander, and Klaus R. Scherer. "That Baby Caught My Eye . . . : Attention Capture by Infant Faces." *Emotion* 7(3) (2007): 685–689.

56 *Conversely, psychologist Teresa Farroni . . .* : Farroni, Teresa, Gergely Csibra, Fran-cesca Simion, and Mark H. Johnson. "Eye Contact Detection in Humans from Birth." *Proceedings of the National Academy of Sciences of the United States of America* 99 (2002): 9602–9605.

56 *To demonstrate, Chris Friesen . . .* : Friesen, Chris K., and Alan Kingstone. "The Eyes Have It! Reflexive Orienting Is Triggered by Nonpredictive Gaze." *Psy-chonomic Bulletin and Review* 5(3) (1998): 490–495.

57 *But what does it tell us . . .* : For those interested in a more detailed discussion of the processes underlying face perception, see Elaine M. Fox and Konstantina Zougkou, "Individual Differences in the Processing of Facial Expressions," in *The*

Handbook of Face Perception, ed. Andrew Calder, Gillian Rhodes, James V. Haxby, and Mark H. Johnson (Oxford: Oxford University Press, 2010).

57 *Back in the 1960s, the social psychologist Stanley Milgram . . .* : Milgram, Stanley, Leonard Bickman, and Lawrence Berkowitz. "Note on the Drawing Power of Crowds of Different Size." *Journal of Personality and Social Psychology* 13 (1969): 79–82.

58 *Most children acquire the rudiments of a Theory of Mind . . .* : Wimmer, Heinz, and Josef Perner. "Beliefs about Beliefs: Representation and Constraining Function of Wrong Beliefs in Young Children's Understanding of Deception." *Cognition* 13 (1983): 103–128.

59 *Disorders of the autistic spectrum . . .* : ToM deficits have also been implicated in schizophrenia and psychopathy, as well as in anorexia and depression, but not to the same extent as in disorders of the autistic spectrum. Similarly, irregularities in eye contact are also found in other disorders (e.g., social anxiety and depression) but again, do not feature as prominently as those found in autism.

62 *Studies have shown . . .* : For more on perceptual contrast and eye coloration see Pawan Sinha, "Here's Looking at You Kid," *Perception* 29 (2000): 1005–1008; Paola Ricciardelli, Gordon Baylis, and Jon Driver, "The Positive and Negative of Human Expertise in Gaze Perception," *Cognition* 77 (2000): B1–B14; and Hiromi Kobayashi and Shiro Kohshima, "Unique Morphology of the Human Eye," *Nature* 387 (1997): 767–768.

63 *"We have an uncanny ability . . ."*: Guthrie, R. D. "Evolution of Human Threat Display Organs." In *Evolutionary Biology,* vol. 4, ed. Dobzhansky, Theodosius, Max K. Hecht, and William C. Steere, 257–302. New York: Appleton Century Crofts, 1970.

3. MIND THEFT AUTO

67 *Sloan, in fact, might well be onto something here . . .* : (Footnote 1) Schauss, Alexander G. "The Physiological Effect of Colour on the Suppression of Human Aggression: Research on Baker-Miller Pink." *International Journal of Biosocial Research* 7(2) (1985): 55–64. For more on the effects of Baker-Miller pink see James E. Gilliam, "The Effects of Baker-Miller Pink on Physiological and Cognitive Behaviours of Emotionally Disturbed and Regular Education Students," *Behavioural Disorders* 17 (1991): 47–55; and Pamela J. Profusek and David W. Rainey, "Effects of Baker-Miller Pink and Red on State Anxiety, Grip Strength and Motor Precision," *Perceptual and Motor Skills* 65 (1987): 941–942.

71 *Take, for example, the two photographs of Margaret Thatcher . . .* : Figure 3.2 taken from Thompson, Peter. "Margaret Thatcher: A New Illusion." *Perception* 9(4) (1980): 483–484. Figure reproduced with permission from Pion Limited, London.

73 *Ellen Langer, professor of psychology . . .* : Langer, Ellen J., Arthur Blank, and Benzion Chanowitz. "The Mindlessness of Ostensibly Thoughtful Action: The Role of 'Placebic' Information in Interpersonal Interaction." *Journal of Personality and Social Psychology* 36 (1978): 635–642.

76 *The principle of "cognitive load" . . .* : For a detailed and accessible account of cog-

266   NOTES

nitive load and visual search studies see Anne M. Treisman, "Features and Objects: The Fourteenth Bartlett Memorial Lecture," *Quarterly Journal of Experimental Psychology* 40A (1988): 201–237.

78 *Imagine that you're working...*: Beyth, Marom Ruth, and Shlomith Dekel. *An Elementary Approach to Thinking under Uncertainty.* Hillsdale, NJ: Erlbaum, 1985.

79 *Heuristics are pretty much...*: For further information on cognitive shortcuts see Daniel Kahneman and Amos Tversky, "On the Psychology of Prediction," *Psychological Review* 80 (1973): 237–251.

80 *In a study which looked at the effects of expectation...*: Plassman, Hilke, John P. O'Doherty, Baba Shiv, and Antonio Rangel. "Marketing Actions Can Modulate Neural Representations of Experienced Pleasantness." *Proceedings of the National Academy of Sciences of the United States of America* 105(3) (2008): 1050–1054.

81 *Similar results have also been found...*: Brochet, Frederic. "Chemical Object Representation in the Field of Consciousness." Working paper (2001): General Oenology Laboratory, France.

81 *John Darley and Paul Gross at Princeton University...*: Darley, John M., and Paul H. Gross. "A Hypothesis Confirming Bias in Labeling Effects." *Journal of Personality and Social Psychology* 44 (1983): 20–33.

82 *Margaret Shih at Harvard...*: Shih, Margaret, Todd L. Pittinsky, and Nalini Ambady. "Stereotype Susceptibility: Identity Salience and Shifts in Quantitative Performance." *Psychological Science* 10 (1999): 80–83.

82 *Jeff Stone, at the University of Arizona...*: Stone, Jeff, Christian I. Lynch, Mike Sjomeling, and John M. Darley. "Stereotype Threat Effects on Black and White Athletic Performance." *Journal of Personality and Social Psychology* 77(6) (1999): 1213–1227.

82 *To illustrate, consider the following...*: Examples taken from Slovic, Paul, Baruch Fischhoff, and Sarah Lichtenstein. "Cognitive Processes and Societal Risk Taking." In *Cognition and Social Behavior,* ed. Carroll, John S., and John W. Payne. Hillsdale, NJ: Erlbaum, 1976.

83 *It's difficult to convey...*: For an accessible account of heuristics see Charles G. Lord, *Social Psychology,* 49–99, chap. 2 (Fort Worth, TX: Harcourt Brace, 1997).

85 *Psychologist David Strohmetz...*: Strohmetz, David B., Bruce Rind, Reed Fisher, and Michael Lynn. "Sweetening the Till: The Use of Candy to Increase Restaurant Tipping." *Journal of Applied Social Psychology* 32 (2002): 300–309.

87 *In 1971, the late Henri Tajfel...*: Tajfel, Henri, Michael G. Billig, Robert P. Bundy, and Claude Flament. "Social Categorization and Intergroup Behaviour." *European Journal of Social Psychology* 1 (1971): 149–178.

88 *In 1955, the American social psychologist Solomon Asch...*: Asch, Solomon E. "Opinions and Social Pressure." *Scientific American* 193 (1955): 31–35.

90 *There are few better examples...*: Fein, Steven, George R. Goethals, Saul M. Kassin, and Jessica Cross. "Social Influence and Presidential Debates." Paper presented at the American Psychological Association convention, 1993.

92 *Exactly the same principle...*: Uhlhaas, Christoph. "Is Greed Good?" *Scientific American Mind,* August/September, 2007.

93 *Ferguson remembers it like this...*: Interview conducted by Glenn Moore (*The Independent,* June 3, 2008).

94 *And those that are not, a recent study shows...*: Grosbras, Marie Helène, Marije

Jansen, Gabriel Leonard, Anthony McIntosh, Katja Osswald, Catherine Poulsen, Laurence Steinberg, Roberto Toro, and Thomas Paus. "Neural Mechanisms of Resistance to Peer Influence in Early Adolescence." *Journal of Neuroscience* 27(30) (2007): 8040–8045.

95 *Young male syndrome . . .* : Buss, David M., and Joshua D. Duntley. "The Evolution of Aggression." In *Evolution and Social Psychology,* ed. Mark Schaller, Douglas T. Kenrick, and Jeffry A. Simpson. New York: Psychology Press, 2006.

95 *"One of the unique dynamics . . .":* Groth, A. Nicholas, and H. Jean Birnbaum. *Men Who Rape: The Psychology of the Offender.* New York: Plenum Press, 1979.

96 *A similar phenomenon . . .* : I first heard about "the gift" or "posing up" in a 2007 BBC television documentary called "HIV and Me" presented by Stephen Fry (www.telegraph.co.uk/culture/tvandradio/3668295/Lastnightontelevision StephenFryHIVandMeBBC2--GreatBritishJourneysBBC2.html). Though not a common practice, my experience in San Francisco corroborates what Fry discovered—that this is a custom found among a small minority of the gay community.

4. PERSUASION GRANDMASTERS

101 *"The order in which you give information . . .":* Lemann, Nicholas. "The Word Lab." *The New Yorker,* October 16, 2000.

103 *Precisely these two scenarios . . .* : Alicke, Mark D. "Culpable Causation." *Journal of Personality and Social Psychology* 63 (1992): 368–378.

104 *Just how fundamental . . .* : Ross, Lee D., Teresa M. Amabile, and Julia L. Steinmetz. "Social Roles, Social Control, and Biases in Social Perception Processes." *Journal of Personality and Social Psychology* 35 (1977): 485–494.

106 *Studies involving mock juries . . .* : Jones, Cathaleene, and Elliot Aronson. "Attributions of Fault to a Rape Victim as a Function of the Respectability of the Victim." *Journal of Personality and Social Psychology* 26 (1973): 415–419; Luginbuhl, James, and Courtney Mullin. "Rape and Responsibility: How and How Much Is the Victim Blamed?" *Sex Roles* 7 (1981): 547–559.

107 *Psychologist George Bizer . . .* : Bizer, George Y., and Richard E. Petty. "How We Conceptualize Our Attitudes Matters: The Effects of Valence Framing on the Resistance of Political Attitudes." *Political Psychology* 26 (2005): 553–568.

109 *In 2006, a team of German psychologists . . .* : Englich, Birte, Thomas Mussweiler, and Fritz Strack. "Playing Dice with Criminal Sentences: The Influence of Irrelevant Anchors on Experts' Judicial Decision Making." *Personality and Social Psychology Bulletin* 32 (2006): 188–200.

109 *Chris Janiszewski and Dan Uy . . .* : Janiszewski, Chris, and Dan Uy. "Precision of Anchor Influences the Amount of Adjustment." *Psychological Science* 19(2) (2008): 121–127.

111 *"When we decide in the blink of an eye . . .":* Meyer, Michelle. "Good Things Come in New Packages." *Arrive,* November/December, 2007.

111 *"They're selling you the experience . . .":* Ibid.

114 *Robert Cialdini, Regents' professor of psychology . . .* : Cialdini, Robert B., Joyce E. Vincent, Stephen K. Lewis, Jose Catalan, Diane Wheeler, and Betty L. Derby. "Re-

ciprocal Concessions Procedure for Inducing Compliance: The Door in the Face Technique." *Journal of Personality and Social Psychology* 31 (1975): 206–215.

115 *Just how much influence . . .* : Cialdini, Robert B. "The Science of Persuasion." *Scientific American Mind,* February, 2001.

116 *Sinclair's technique has a name . . .* : Freedman, Jonathan L., and Scott C. Fraser. "Compliance without Pressure: The Foot in the Door Technique." *Journal of Personality and Social Psychology* 4 (1966): 195–203.

117 *In sales, a related technique . . .* : For more on low-balling see Robert Cialdini, *Influence: Science and Practice,* 4th ed. (Boston: Allyn and Bacon, 2001).

119 *In 1946, Solomon Asch . . .* : Asch, Solomon E. "Forming Impressions of Personality." *Journal of Abnormal and Social Psychology* 41 (1946): 258–290.

121 *In 2005, the Global Language Monitor . . .* : For a lighthearted tour of the latest politically (in)correct words and phrases see www.languagemonitor.com/news/top-politically-incorrect-words-of-2009; for the original "misguided criminals" article see John Simpson, "London Bombs Need Calm Response," BBC Home, August 31, 2005, http://news.bbc.co.uk/1/hi/uk/4671577.stm (accessed November 17, 2005).

122 *A classic study conducted in 1974 . . .* : Loftus, Elizabeth F., and John C. Palmer. "Reconstruction of Automobile Destruction: An Example of the Interaction between Language and Memory." *Journal of Verbal Learning and Verbal Behaviour* 13 (1974): 585–589.

123 *"As soon as the race label is added . . ."*: Von Drehle, David. "Five Faces of Obama." *Time,* August 21, 2008.

124 *Back in 2000,* New Yorker *correspondent . . .* : Lemann, Nicholas. "The Word Lab." *The New Yorker,* October 16, 2000.

5. PERSUASION BY NUMBERS

130 *Just how easy it is . . .* : For more on the phenomenon of group polarization see Rupert Brown, *Group Processes,* 142–158 (Oxford: Blackwell, 1993).

131 *This is something you can demonstrate . . .* : Wallach, Michael A., Nathan Kogan, and Daryl J. Bem. "Group Influence on Individual Risk Taking." *Journal of Abnormal and Social Psychology* 65 (1962): 75–86.

131 *The effects of group polarization . . .* : For a more detailed look at how the decision-making process varies between individuals and groups, see Cass R. Sunstein, *Going to Extremes: How Like Minds Unite and Divide* (New York: Oxford University Press, 2009).

132 *Research has shown . . .* : Myers, David G., and George D. Bishop. "Discussion Effects on Racial Attitudes." *Science* 169 (1970): 778–779.

133 *These, laboratory studies have shown . . .* : For a review of the factors that both increase and reduce conformity, see Elliot Aronson, *The Social Animal,* 5th ed., chap. 2 (New York: W. H. Freeman & Company, 1988).

134 *In 2007, he and his colleagues . . .* : Goldstein, Noah J., Robert B. Cialdini, and Vladas Griskevicius. "A Room with a Viewpoint: Using Social Norms to Motivate Environmental Conservation in Hotels." *Journal of Consumer Research* 35 (2008):

472–482; Goldstein, Noah J., Robert B. Cialdini, and Vladas Griskevicius. "Invoking Social Norms: A Social Psychology Perspective on Improving Hotels' Linen-Reuse Programs." *Cornell Hotel and Restaurant Administration Quarterly,* May, 2007, www.entrepreneur.com/tradejournals/article/163394867_2.html (accessed September 24, 2009).

135 *In 1980, the French social psychologist . . .* : Moscovici, Serge, and Bernard Personnaz. "Studies in Social Influence: V. Minority Influence and Conversion Behaviour in a Perceptual Task." *Journal of Experimental Social Psychology* 16 (1980): 270–282.

136 *Trouble is . . .* : For a detailed discussion of the ups and downs of Moscovici's "afterimage paradigm," see Robin Martin, "Majority and Minority Influence Using the Afterimage Paradigm: A Series of Attempted Replications," *Journal of Experimental Social Psychology* 34(1) (1998): 1–26.

136 *The study may be split . . .* : The experiment actually consisted of four phases (see original paper, detailed in previous note) but in the interests of clarity I have collapsed these into two.

139 *In Figure 5.3, you have four cards . . .* : Wason, Peter C. "Reasoning." In *New Horizons in Psychology,* ed. Brian M. Foss, 135–151. Harmondsworth: Penguin, 1966. For more on the Wason selection task and hypothesis testing in general, see Alan Garnham and Jane Oakhill, *Thinking and Reasoning,* chap. 8 (Oxford: Blackwell, 1994).

141 *Back in 1979, psychologists . . .* : Snyder, Mark, and Nancy Cantor. "Testing Hypotheses about Other People: The Use of Historical Knowledge." *Journal of Experimental Social Psychology* 15 (1979) 330–342.

141 *In an amusing, ingenious . . .* : Henderson, Charles E. "Placebo Effects Prove the Value of Suggestion." www.biocentrix.com/hypnosis/placebo.htm (accessed May 28, 2009).

144 *In fact a recent study . . .* : Wiltermuth, Scott S., and Chip Heath, "Synchrony and Cooperation." *Psychological Science* 20 (2009): 1–5.

144 *Social psychologist Miles Hewstone . . .* : Islam, Mir R., and Miles Hewstone, "Intergroup Attributions and Affective Consequences in Majority and Minority Groups." *Journal of Personality and Social Psychology* 64 (1993): 936–950.

145 *Research has shown that . . .* : Miller, Richard L., Philip Brickman, and Diana Bolen, "Attribution versus Persuasion as a Means for Modifying Behavior." *Journal of Personality and Social Psychology* 31 (1975): 430–441.

146 *Take a condition . . .* : For an easy-to-read introduction to the Stockholm syndrome see Joseph M. Carver, "Love and Stockholm Syndrome: The Mystery of Loving an Abuser," Counseling Resource, http://counsellingresource.com/quizzes/stockholm/index.html (accessed November 20, 2009).

147 *Are we really surprised . . .* : A number of observations allude to the possibility that Natascha Kampusch was suffering from Stockholm syndrome. According to police, she cried inconsolably when told that Wolfgang Priklopil was dead, and lit a candle for him in the mortuary. "My youth was very different," she has said of her time in captivity. "But I was also spared a lot of things. I did not start smoking or drinking, and I did not hang out in bad company." (Julia Layton, "What Causes Stockholm Syndrome?" How Stuff Works, http://health.howstuffworks

.com/mental-health/mental-disorders/stockholm-syndrome.htm [accessed December 14, 2009].) Bizarrely (even by her own admission), Kampusch now *owns* the house in which Priklopil kept her prisoner — claiming it from his estate to prevent its demolition. "I know it's grotesque," she acknowledges. "I must now pay for electricity, water and taxes on a house I never wanted to live in." ("Kidnap Victim Owns Her House of Horrors," Sky News, May 15, 2008, http://news.sky .com/skynews/Home/SkyNewsArchive/Article/20080641316125 [accessed May 23, 2008].) Indeed, Kampusch has visited the site regularly — fueling speculation that she might one day move back into it. For more on the extraordinary case of Natascha Kampusch, see Bojan Pancevski and Stefanie Marsh, "Natascha Kampusch: From Darkness to Limelight," Times Online, June 2, 2008, http://women .timesonline.co.uk/tol/life_and_style/women/article4044283.ece (accessed August 30, 2008).

147 *In the mid-1960s, cognitive psychologist Martin Seligman . . .* : Seligman, Martin E. P., and Steven F. Maier, "Failure to Escape Traumatic Shock." *Journal of Experimental Psychology* 74 (1967): 1–9.

148 *Though he has since repudiated . . .* : For more on the role of psychology in the development of interrogation techniques, see Jane Mayer, *The Dark Side: The Inside Story of How the War on Terror Turned into a War on American Ideals* (New York: Doubleday, 2008).

148 *It depends on your attributional style . . .* : For more on attributional, or explanatory, style see Martin E. P. Seligman, *Learned Optimism: How to Change Your Mind and Your Life* (New York: Random House, 2006).

150 *A study in the 1970s . . .* : Glass, David C., and Jerome E. Singer. *Urban Stress: Experiments on Noise and Urban Stressors.* New York: Academic Press, 1972.

152 *The taxonomy that . . .* : For more information on the typology of domestic abuse, see Pat Craven, *The Freedom Programme* (2005), www.freedomprogramme.co.uk.

6. SPLIT-SECOND PERSUASION

158 *One afternoon, in a classroom . . .* : This anecdote crops up in most biographies of Gauss.

162 *Consider, for example . . .* : Eastaway, Rob, and Jeremy Wyndham. *Why Do Buses Come in Threes? The Hidden Mathematics of Everyday Life.* London: Robson Books, 1998.

163 *Luke Conway, professor of psychology . . .* : Thoemmes, Felix, and Lucian. G. Conway III. "Integrative Complexity of 41 U.S. Presidents." *Political Psychology* 28 (2007): 193–226.

164 *Matthew McGlone . . . and Jessica Tofighbakhsh . . .* : McGlone, Matthew S., and J. Tofighbakhsh. "Birds of a Feather Flock Conjointly: Rhyme as Reason in Aphorisms." *Psychological Science* 11 (2000): 424–428.

167 *And making sure they feel good about it . . .* : Usually, the boot is on the other foot and it's the adult scrabbling around for things that are in the *child's* self-interest to wangle some peace and quiet (PlayStation, chocolate, and hard cash figuring among the preferred currencies of persuasion in such cases). In fact, receipt of re-

ward and avoidance of punishment comprise the twin pillars of pretty much any kind of influence you can think of, from childrearing, through puppy training, to the more esoteric and Machiavellian endpoints of consciously shaped behavior such as that found in sniffer dogs and, very nearly, in pigeons during the Second World War.

Project Pigeon was the brainchild of American psychologist B. F. Skinner and constituted the code name for a scheme aimed at developing a pigeon-guided missile. The idea, if eccentric, was simple. A lens affixed to the front of the missile would project an image of the target onto an interior screen, while a pigeon, previously trained through operant conditioning to recognize it (i.e., rewarded with food for incrementally accurate differentiation) pecked at its adjustable surface. So long as the pecks were directed center screen the missile would remain on course. But pecks directed off center would cause the screen to tilt, resulting, via a link to the missile's control system, in the flight path deviating accordingly.

Unfortunately, despite an initial outlay of $25,000 by the U.S. National Defense Research Committee, and encouraging results early on, the project, so to speak, never got off the ground. Bird brains and missile defense, the committee concluded, were a bad combination.

Conditioning, of course, doesn't always have to take place consciously. Sometimes, unlike with the pigeons, self-interest can be manipulated without our even knowing, and the association between performing a certain behavior (pecking) and achieving a given outcome (food) forged implicitly.

Katie: Mum, can I have an ice lolly?

Mum: (Takes no notice and continues lying on sun lounger.)

Katie: Mum, I want an ice lolly.

Mum: Yes, in a minute, darling.

Katie: (Stamps her foot.) MUM, I WANT AN ICE LOLLY!

Mum: Hey! How many times have I told you before? Don't raise your voice to me. Now, which one do you want?

Little Katie doesn't need a Ph.D. in psychology, and an in-depth knowledge of the workings of the human mind, to know that increasingly importunate persistence often pays off. She knows from experience: she's been honing her skills right from the day she was born! (To back her up, psychologist Edward Burkley of Oklahoma State University studied the impact of cognitive fatigue on resistance to persuasion in seventy-eight students. He found that students who were tired were more likely to accept a two-month cut in holiday over the summer than those who were fresh.)

But persistence isn't *always* a means to an end. It can, on occasion, constitute the end itself: the behavior that's being rewarded. This peculiar dynamic is the psychology behind gambling and fishing, and is what makes both these ancient pastimes so difficult to give up. Roulette wheels and riverbanks implement variable, unpredictable reward contingencies that play on the addictiveness of hope (you never know when the next big payout is coming), and it is *this* — hope's capacity to spring eternal, as opposed to the impromptu catch, or sudden, fortuitous hot streak — that unwittingly, over time, reels one in.

For an accessible introduction to the principles of operant conditioning, and

an easy-to-read overview of how schedules of reward and punishment can rein-
force behavior, see David G. Myers, *Psychology,* 4th ed., chap. 8, 257–285 (New
York: Worth, 1995). For Edward Burkley's study see "The Role of Self-control in
Resistance to Persuasion," *Personality and Social Psychology Bulletin* 34(3) (2008):
419–431.

167 *Scarcity is one of six . . .* : For a good introduction to these six principles of persuasion see Robert B. Cialdini, "The Science of Persuasion," *Scientific American
Mind,* February 2001.

167 *A recent study from the University of Aberdeen . . .* : Jones, Benedict C., Lisa M.
DeBruine, Anthony C. Little, Robert P. Burriss, and David R. Feinberg. "Social
Transmission of Face Preferences among Humans." *Proceedings of the Royal Soci-
ety of London* B 274(1611) (2007): 899–903.

169 *Psychologists John Darley and Daniel Batson . . .* : Darley, John M., and C. Daniel Batson. "From Jerusalem to Jericho: A Study of Situational and Dispositional
Variables in Helping Behaviour." *Journal of Personality and Social Psychology* 27
(1973): 100–108.

171 *You don't find many conjuring tricks . . .* : www.hondomagic.com/html/a_little_
magic.htm.

173 *Magicians, of course . . .* : For more on what magic can teach us about cognitive
processes see Gustav Kuhn, Alym A. Amlani, and Ronald A. Rensink, "Towards a
Science of Magic," *Trends in Cognitive Sciences* 12(9) (2008): 349–354.

174 *The mercurial British barrister . . .* : www.anecdotage.com (accessed July 3, 2007).

176 *This paradigm — a variant . . .* : For a scholarly review of the Stroop effect literature see Colin M. MacLeod, "Half a Century of Research on the Stroop Effect: An
Integrative Review," *Psychological Bulletin* 109(2) (1991): 163–203.

176 *Barbara Davis and Eric Knowles . . .* : Davis, Barbara P., and Eric S. Knowles. "A
Disrupt Then Reframe Technique of Social Influence." *Journal of Personality and
Social Psychology* 76 (1999): 192–199.

177 *Single-cell recordings in monkeys . . .* : Belova, Marina A., Joseph J. Paton, Sara
E. Morrison, and C. Daniel Salzman. "Expectation Modulates Neural Responses
to Pleasant and Aversive Stimuli in Primate Amygdala." *Neuron* 55 (2007): 970–
984.

177 *While in humans . . .* : Halgren, Eric, and Ksenija Marinkovic. "Neurophysiologi-
cal Networks Integrating Human Emotions." In *Cognitive Neuroscience,* ed. Michael S. Gazzaniga, 1137–1151. Cambridge, MA: MIT Press, 1995.

177 *And gives music and humor . . .* : For a detailed (and entertaining) discussion of
the anatomy of humor, see Jimmy Carr and Lucy Greeves, *The Naked Jape: Uncovering the Hidden World of Jokes* (London: Penguin, 2007).

177 *Take, for instance, the two advertisements . . .* : "Here to Help" poster reproduced
with permission of Network Rail ©. "We Try Harder."

178 *Westen and his coworkers . . .* : Westen, Drew, Pavel S. Blagov, Clint Kilts, Keith
Harenski, and Stephan Hamann. "Neural Bases of Motivated Reasoning: An fMRI
Study of Emotional Constraints on Partisan Political Judgement in the 2004 U.S.
Presidential Election." *Journal of Cognitive Neuroscience* 18 (2006): 1947–1958.

179 *Facts aren't always that important . . .* : Attitude change researchers have demonstrated precisely when facts are important, and when they are not, in persuasion.

Generally, if an attitude is primarily "cognitive" in nature (i.e., emerges from how we *think* about something), then logical, rational argument by far and away constitutes the most effective tool in bringing about change. If, on the other hand, an attitude is primarily "affective" in nature (i.e., has to do with how we *feel* about something), then an appeal to the emotions offers the best strategy. This might sound complicated, but it is actually something that most of us already know. Think, for instance, of the advertising industry. Anyone who attempts to sell perfume by extolling the virtues of its chemical composition, or drain unblocker by its power to make one irresistible to members of the opposite sex, is not, in all honesty, going to be the next Bill Gates.

Closely linked to the content of a persuasive message is the way in which we, as recipients, process the information within it. As with message composition, there are two main routes to persuasion here, the *central route* and *peripheral route* (see figure), the critical differential this time residing less in the manner in which we reach our conclusions (whether we use our head or our heart), and more in our motives for doing so.

Two routes to persuasion: central versus peripheral.

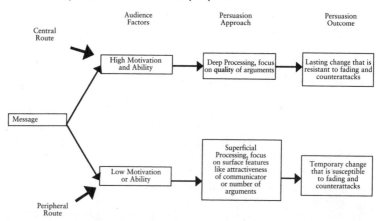

Broadly speaking, we tend to process information via the central route when it is of high personal relevance to us: in other words, when it really matters. Normally, this entails an in-depth appraisal of the quality of specific arguments, and results in lasting attitude change. In contrast, the peripheral route is activated under conditions of low personal involvement: when the stakes are not so high. Peripheral processing of a message or argument is characterized by a reduced focus on details and an enhanced reliance on superficial factors such as the physical attractiveness of the communicators, or how they are dressed. Thus, just as we wouldn't recruit a biochemist to market a perfume, or a supermodel to sell drain unblocker, we'd be similarly ill-advised to choose our mortgage broker purely on the strength of his Elvis impersonation, or, alternatively, insist on a full explication of the laws of quantum physics before rolling the dice in a game of Trivial Pursuit.

Recall from Chapter 2 how good-looking individuals make more money on charity stands than those of average looks? Now you know why. Most of us already harbor the notion that giving to charity is good, and, from a strictly intellectual point of view (the kind handled by the central route of persuasion) need no further convincing. Rather, cue peripheral route, we need *inducing*.

For more on the science of attitude change, see William D. Crano and Radmila Prislin, "Attitudes and Persuasion," *Annual Review of Psychology* 57 (2006): 345–374.

179 *In fact, research has shown . . .* : Greatbatch, David, and John Heritage. "Generating Applause: A Study of Rhetoric and Response at Party Political Conferences." *The American Journal of Sociology* 92 (1986): 110–157.

182 *Or take Stanley Milgram's electric shocks experiment . . .* : In 1963, Milgram published a study that has now assumed iconic status in the field of experimental psychology — and which has arguably gone down as the most celebrated, certainly the most combustible, in the discipline's hundred or so year history. Milgram devised a simulated learning paradigm in which participants (recruited from a random sample of respectable, middle-class Americans) were allocated the role of "teacher" opposite an associate of the researcher (the "learner"). But this was no ordinary teaching assignment. "Mistakes" were punished by the administration of electric shocks — minimal at the outset but escalating to a brutal 450 volts as errors perpetuated. Ostensibly, the study was presented as an investigation into short-term memory. And ostensibly, the electric shocks were real. But the *actual* focus was on obedience — and the shocks were a sham. The aim was chillingly simple: To what extremes, Milgram wanted to know, were everyday, law-abiding American citizens prepared to go when instructed by an authority figure? For more on Milgram's research into obedience, see his book *Obedience to Authority: An Experimental View* (New York: HarperCollins, 1974).

182 *A recent study . . .* : McCabe, David P., and Alan D. Castel. "Seeing Is Believing: The Effect of Brain Images on Judgements of Scientific Reasoning." *Cognition* 107 (2008): 343–352.

183 *Statistics, used well . . .* : Leonard Mlodinow. *The Drunkard's Walk: How Randomness Rules Our Lives.* New York: Pantheon Books, 2008.

183 *Psychologist Paul Zarnoth . . .* : Zarnoth, Paul, and Janet A. Sniezek. "The Social Influence of Confidence in Group Decision Making." *Journal of Experimental Social Psychology* 33 (1996): 345–366.

184 *(Footnote) The extent to which nurses . . .* : For more on nurses and the inhibition of facial expression see Wallace P. Ekman, Wallace V. Friesen, and Maureen O'Sullivan, "Smiles When Lying," *Journal of Personality and Social Psychology* 54(3) (1988): 414–420.

184 *Psychologists Nalini Ambady . . .* : Ambady, Nalini, and Robert Rosenthal. "Half a Minute: Predicting Teacher Evaluations from Thin Slices of Nonverbal Behavior and Physical Attractiveness." *Journal of Personality and Social Psychology* 64(3) (1993): 431–441.

186 *Lisa DeBruine at the University of Aberdeen's . . .* : DeBruine, Lisa M. "Facial Resemblance Enhances Trust." *Proceedings of the Royal Society of London* B 269 (2002): 1307–1312.

186 *First, DeBruine devised a computer game . . .* : DeBruine's computer task is a variation of a well-known paradigm within game theory called the ultimatum game. For more on the ultimatum game, see Steven D. Levitt and Stephen J. Dubner, *Super Freakonomics: Global Cooling, Patriotic Prostitutes, and Why Suicide Bombers Should Buy Life Insurance* (New York: HarperCollins, 2009).

187 *Would they, as predicted by . . .* : In ethology, kin selection refers to the tendency for some animals to favor others that bear most genetic similarity to themselves. For those interested in learning more about kin selection, the following articles offer an excellent grounding in the subject. Hamilton, William D. "The Evolution of Altruistic Behavior." *American Naturalist* 97 (1963): 354–356; Hamilton, William D. "The Genetical Evolution of Social Behavior." *Journal of Theoretical Biology* 7(1) (1964): 1–52; Smith, J. Maynard. "Group Selection and Kin Selection." *Nature* 201(4924) (1964): 1145–1147.

188 *An amusing study . . .* : Finch, John F., and Robert B. Cialdini. "Another Indirect Tactic of (Self) Image Management: Boosting." *Personality and Social Psychology Bulletin* 15 (1989): 222–232.

188 *By not referring to . . .* : For an accessible amalgamation of scientific research and the techniques of classical rhetoric see Max Atkinson, *Lend Me Your Ears: All You Need to Know about Making Speeches and Presentations* (London: Vermillion, 2004).

189 *I have absolutely no doubt . . .* : For more on mind-reading geniuses see Malcolm Gladwell, "The Naked Face," *The New Yorker* Archive (August 5, 2002), www .gladwell.com/2002/2002_08_05_a_face.htm (accessed June 11, 2008).

190 *The importance of nailing . . .* : Ottati, Victor, Susan Rhoads, and Arthur C. Graesser. "The Effect of Metaphor on Processing Style in a Persuasion Task: A Motivational Resonance Model." *Journal of Personality and Social Psychology* 77(4) (1999): 688–697. Related to the concept of nailing the right frequency is the common-sense idea that persuasion becomes increasingly more likely the less a person has to change his or her mind. Back in the early 1960s, the social psychologists Muzafer Sherif and Carl Hovland put forward the notion that attitudes, like earthquakes, have epicenters — and that the greater the distance between the "eye" of a person's belief and the fulcrum of an influence attempt, the less likely he or she is to be swayed. The key, according to Sherif and Hovland, is to pitch a message within the target's "latitude of acceptance" — in other words, to make sure that whatever it is you're proposing isn't so "off the scale" as to be rejected outright. Imagine, for instance, that you're trying to persuade a friend to lose weight. Do you think you'd stand a better chance if you signed him up for a marathon — or extolled the benefits of a gentle walk round the block?

Framing, needless to say, is often the persuader's sneaky secret weapon when it comes to latitudes of acceptance. An example occurred with a friend of mine — a macho, muscle-bound guy whose anger-management issues were threatening to wreck his marriage. The thought of taking time out to meditate was a total no-no to him — the preserve of counselors, hippies, and other (as he put it) "middle-class puffs." Then one day he made a discovery: meditation, as well as being practiced by dreadlocked suburban yogis, was also endemic to many of the martial arts. That, of course, was different. Suddenly, the idea of taking time out wasn't so namby-pamby after all — and he embarked on a regular routine.

For the original work on latitudes of acceptance and rejection, see Muzafer Sherif and Carl Hovland, *Social Judgment: Assimilation and Contrast Effects in Communication and Attitude Change* (New Haven: Yale University Press, 1961).

191 *Recently, the CIA* . . . : Hardcastle, Ephraim. Mail Online. www.dailymail.co.uk/debate/article1213906/EPHRAIMHARDCASTLE.html (accessed October 19, 2009).

191 *Research has shown* . . . : Carlson, Michael, Ventura Charlin, and Norman Miller. "Positive Mood and Helping Behavior: A Test of Six Hypotheses." *Journal of Personality and Social Psychology* 55 (1988): 211–229.

192 *Research, for example* . . . : Winkielman, Piotr, and John T. Cacioppo. "Mind at Ease Puts Smile on the Face: Psychophysiological Evidence That Processing Facilitation Elicits Positive Affect." *Journal of Personality and Social Psychology* 81(6) (2001): 989–1000.

7. THE PSYCHOPATH — NATURAL BORN PERSUADER

199 *The Psychopathy Checklist Revised (PCL-R)* . . . : Hare, Robert D. *The Hare Psychopathy Checklist — Revised (PCL-R),* 2nd ed. Toronto, Ontario: MultiHealth Systems, 2003.

200 *Psychologists Scott Lilienfeld and Brian Andrews* . . . : Lilienfeld, Scott O., and Brian P. Andrews. "Development and Preliminary Validation of a Self-Report Measure of Psychopathic Personality in Noncriminal Populations." *Journal of Personality Assessment* 66 (1996): 488–524.

202 *There are two types* . . . : For a scholarly and accessible introduction to empathy see Mark H. Davis, *Empathy: Asocial Psychological Approach* (New York: HarperCollins, 1996). For more on differences between hot and cold empathy see George Loewenstein, "Hot Cold Empathy Gaps and Medical Decision Making," *HealthPsychology* 24(4) (2005): S49–S56; and Daniel Read and George Loewenstein, "Enduring Pain for Money: Decisions Based on the Perception and Memory of Pain," *Journal of Behavioral Decision Making* 12 (1999): 1–17.

202 *The comparison between psychopaths and nonpsychopaths* . . . : For an in-depth discussion of brain dysfunction, emotional processing deficits, and psychopathy (including moral dilemmas) see R. J. R. Blair, "Dysfunctions of Medial and Lateral Orbitofrontal Cortex in Psychopathy," *Annals of the New York Academy of Sciences* 1121 (2007): 461–479. For a less specialized account see Carl Zimmer, "Whose Life Would You Save?" *Discover* (April 2004), http://discovermagazine.com/2004/apr/whose-life-would-you-save (accessed January 9, 2007).

202 *Consider, for example* . . . : The trolley problem was first proposed in this form by Philippa Foot in "The Problem of Abortion and the Doctrine of the Double Effect," in *Virtues and Vices and Other Essays in Moral Philosophy* (Berkeley, CA: University of California Press, 1978).

203 *But consider the following* . . . : Thomson, Judith J. "Killing, Letting Die, and the Trolley Problem." *The Monist* 59 (1976): 204–217. Want to take things a stage further? How about this? *A brilliant transplant surgeon has five patients. Each of the patients is in need of a different organ, and each of them will die without that organ.*

*Unfortunately, there are no organs currently available to perform any of the trans-
plants. A healthy young traveler, just passing through, comes into the doctor's surgery for a routine checkup. While performing the checkup, the doctor discovers that
his organs are compatible with all five of his dying patients. Suppose further that
were the young man to disappear, no one would suspect the doctor.* . . . (Thomson,
Judith J. "The Trolley Problem." *Yale Law Journal* 94 (1985): 1395–1415.)

203 *Harvard psychologist Joshua Greene* . . . : Greene, Joshua D., R. Brian Sommerville,
Leigh E. Nystrom, John M. Darley, and Jonathan D. Cohen. "An fMRI Investiga-
tion of Emotional Engagement in Moral Judgement." *Science* 293 (2001): 2105–
2108. For a more general account of the neuroscience of morality see Joshua D.
Greene and Jonathan Haidt, "How (and Where) Does Moral Judgement Work?"
Trends in Cognitive Sciences 6(12) (2002): 517–523.

204 *But in psychopaths* . . . : An interesting question is whether or not there are also
individuals who score extremely highly on just "hot" empathy — "antipsycho-
paths," in other words. There's evidence to suggest that there might be. Neuroscientist Richard Davidson at the University of Wisconsin, with the help of the Dalai
Lama, has been studying what goes on inside the brains of Buddhist monks — the
Olympic athletes of the meditation world, as he calls them — while they perform
an advanced kind of meditation called "compassionate" meditation. Using EEG,
Davidson has found that when the monks enter into an intensely compassionate
state their heightened focus on unconditional love is accompanied by a unique
neural signature — which includes gamma waves thirty times stronger than normal, and increased activity in the left prefrontal cortex (the part of the brain responsible for positive emotions). Results such as these, Davidson maintains, have
important implications for ongoing research into "neuroplasticity": the ability of
an individual to change brain function through training. Just as the part of the
brain that corresponds to a violinist's fingering hand develops more than the part
that controls the bow hand, so Davidson proposes that the right kind of "training" can also extend to the emotion centers of the brain — and we can bulk up on
empathy just like we can on any other "muscle." (For more on Richard Davidson's
work see the website for the Lab for Affective Neuroscience at the University of
Wisconsin: http://psyphz. psych.wisc.edu/.)

In his book *The Wisdom of Forgiveness,* the Dalai Lama tells the story of
Lopon-la, a Tibetan monk whom he had known in Lhasa before the Chinese in-
vasion — and as good a candidate as any for the title of "antipsychopath." Lopon-la
was imprisoned for eighteen years by the Chinese, and then, on his release,
fled to India. Twenty years after his ordeal, he and the Dalai Lama were finally
reunited.

"He seemed the same," the Dalai Lama recounts. "His mind still sharp after so
many years in prison. He was still the same gentle monk. . . . They tortured him
many times in prison. I asked him whether he was ever afraid. Lopon-la then told
me, 'Yes, there was one thing I was afraid of. I was afraid I may lose compassion
for the Chinese.'"

204 *Similar work* . . . : Gordon, Heather L., Abigail A. Baird, and Alison End. "Functional Differences among Those High and Low on a Trait Measure of Psychopathy." *Biological Psychiatry* 56 (2004): 516–521.

205   *Simon Baron-Cohen, a psychologist* . . . : Richell, R. A., D. G. V. Mitchell, C. New-
         man, A. Leonard, S. Baron-Cohen, and R. J. R. Blair. "Theory of Mind and Psy-
         chopathy: Can Psychopathic Individuals Read the 'Language of the Eyes'?" *Neu-
         ropsychologia* 41 (2003): 523–526.

208   *A 2005 study* . . . : Shiv, Baba, George Loewenstein, Antoine Bechara, Hanna
         Damasio, and Antonio R. Damasio. "Investment Behaviour and the Negative Side
         of Emotion." *Psychological Science* 16(6) (2005): 435–439.

211   *Rachman's studies* . . . : For more on Rachman's work see Stanley J. Rachman,
         "Fear and Courage: A Psychological Perspective," *Social Research* 71(1) (2004):
         149–176. Rachman makes it quite clear in this paper that bomb-disposal experts
         in general are *not* psychopathic — a view echoed here. Rather, the point being
         made is that confidence and coolness under pressure are two traits that psycho-
         paths and bomb-disposal experts have in common.

211   *Robert Hendy-Freegard is* . . . : I should point out that I have not met Robert
         Hendy-Freegard in person, and so cannot say for certain whether or not he is a
         psychopath. His exploits, however, together with the testimonies of both his vic-
         tims and senior police officers, strongly suggest the presence of psychopathic per-
         sonality disorder — and, moreover, that he is pretty high up the scale.

213   *"You may have spotted them* . . .": Baines, David. "The Dark Side of Charisma"
         (book review). *Canadian Business*, May/June 2006.

219   *A few years ago* . . . : Scerbo, Angela, Adrian Raine, Mary O'Brien, Cheryl-Jean
         Chan, Cathy Rhee, and Norine Smiley. "Reward Dominance and Passive Avoid-
         ance Learning in Adolescent Psychopaths." *Journal of Abnormal Child Psychology*
         18(4) (1990): 451–463.

8. HORIZONS OF INFLUENCE

227   *I first ran into the Mirror Man* . . . : Max Coltheart and his colleagues have writ-
         ten extensively on the Mirror Man and delusions of misidentification. To find out
         more see Nora Breen, Diana Caine, and Max Coltheart, "Mirrored-Self Misiden-
         tification: Two Cases of Focal Onset Dementia," *Neurocase* 7 (2001): 239–254;
         and Nora Breen, Diana Caine, Max Coltheart, Julie Hendy, and Corrine Roberts,
         "Towards an Understanding of Delusions of Misidentification: Four Case Stud-
         ies," *Mind and Language* 15(1) (2000): 74–110.

229   *In the summer of 2008* . . . : Marx, David M., Sei Jin Ko, and Ray A. Friedman. "The
         'Obama Effect': How a Salient Role Model Reduces Race-Based Performance Dif-
         ferences." *Journal of Experimental Social Psychology* 45(4) (2009): 953–956.

229   *Back in the 1990s* . . . : For previous work on stereotype threat and race see Claude
         M. Steele, "A Threat in the Air: How Stereotypes Shape Intellectual Identity and
         Performance," *American Psychologist* 52 (1997): 613–629; Claude M. Steele and
         Joshua Aronson, "Stereotype Vulnerability and the Intellectual Test Performance
         of African Americans," *Journal of Personality and Social Psychology* 69 (1995):
         797–811.

231   *Cognitive psychologist Carol Dweck* . . . : For more on Carol Dweck's work on
         mind-sets see Carol S. Dweck, *Mindset: The New Psychology of Success* (New

York: Random House, 2006); and Carol S. Dweck, "The Secret to Raising Smart Kids," *Scientific American Mind,* Dec./Jan. 2007, 36–43.

233 *Kids who take it all in their stride* . . . : For more on the science behind the development of temperament, see Jerome Kagan, *Galen's Prophecy: Temperament in Human Nature* (New York: Basic Books, 1994).

233 *Harvard psychologist Dan Gilbert* . . . : Gilbert, Daniel T., Romin W. Tafarodi, and Patrick S. Malone. "You Can't Unbelieve Everything You Read." *Journal of Personality and Social Psychology* 65(2) (1993): 221–233.

237 *In 1956, the Stanford social psychologist* . . . : To hear the full story of Festinger's escapades with Marion Keech's doomsday group see Leon Festinger, Henry W. Riecken, and Stanley Schachter, *When Prophecy Fails: A Social and Psychological Study of a Modern Group That Predicted the End of the World* (Minneapolis: University of Minnesota Press, 1956).

238 *Festinger's exposé of* . . . : Festinger, Leon. *A Theory of Cognitive Dissonance,* Stanford, CA: Stanford University Press, 1957. See also Leon Festinger and James M. Carlsmith, "Cognitive Consequences of Forced Compliance," *Journal of Abnormal and Social Psychology* 58 (2) (1959): 203–210. For an overview of dissonance theory and competing theories of cognitive consistency see Joel Cooper and Russell H. Fazio, "A New Look at Dissonance Theory," in *Advances in Experimental Social Psychology,* 17, ed. Leonard Berkowitz, 229–266 (Orlando, FL: Academic Press, 1984).

240 *But a recent experiment* . . . : Harris, Sam, Sameer A. Sheth, and Mark S. Cohen. "Functional Neuroimaging of Belief, Disbelief, and Uncertainty." *Annals of Neurology* 63(2) (2008): 141–147. For a comprehensive review of the cognitive bias modification literature see Colin MacLeod, Ernst H. W. Koster, and Elaine Fox, "Whither Cognitive Bias Modification Research? Commentary on the Special Section Articles," *Journal of Abnormal Psychology* 118 (2009): 89–99.

245 *MacLeod has demonstrated this* . . . : MacLeod, Colin, A. Mathews, and Philip Tata. "Attentional Bias in Emotional Disorders." *Journal of Abnormal Psychology* 95 (1986): 15–20.

246 *Recently, MacLeod has been thinking* . . . : MacLeod, Colin, Elizabeth M. Rutherford, Lyn Campbell, Greg Ebsworthy, and Lin Holker. "Selective Attention and Emotional Vulnerability: Assessing the Causal Basis of Their Association through the Experimental Manipulation of Attentional Bias." *Journal of Abnormal Psychology* 111 (2002): 107–123.

246 *And the story doesn't end there* . . . : Mathews, Andrew, and Bundy Mackintosh. "Induced Emotional Interpretation Bias and Anxiety." *Journal of Abnormal Psychology* 109 (2000): 602–615.

248 *In 2004, a couple of years after* . . . : MacLeod, Colin, Lyn Campbell, Elizabeth M. Rutherford, and Edward J. Wilson, "The Causal Status of Anxiety Linked Attentional and Interpretive Bias." In *Cognition, Emotion, and Psychopathology: Theoretical, Empirical, and Clinical Directions,* ed. Jenny Yiend. Cambridge: Cambridge University Press, 2004.

248 *A year later* . . . : Field, Matt, and Brian Eastwood. "Experimental Manipulation of Attentional Bias Increases the Motivation to Drink Alcohol." *Psychopharmacology* 183 (2005): 350–357.

248 *Edward Taub, at the University of Alabama* . . . : For more on Edward Taub and the work of the Taub Therapy Clinic, see Norman Doidge, *The Brain That Changes Itself: Stories of Personal Triumph from the Frontiers of Science,* chap. 5 (New York: Viking, 2007).

252 *Multidimensional Iowa Suggestibility Scale (MISS)* . . . : Kotov, R. I., S. B. Bellman, and D. B. Watson. "Multidimensional Iowa Suggestibility Scale: Brief Manual (2007)." www.stonybrookmedicalcenter.org/sbumcfiles/MISS_FINAL_BLANK_0.pdf and www.stonybrookmedicalcenter.org/sbumcfiles/MISSBriefManual.pdf, with permission from Roman Kotov.

Acknowledgments

Of the three men who feature in the opening section of this book, only one is still alive. My father, John Dutton, called it a day in the spring of 2001, and my friend, the Big Man, threw in his hand less than a year later—on New Year's Day 2002. Guys, this book is written in your memory, and goes out with a message. If you're up there, somewhere, anywhere—let's do it all again sometime.

The bar that I mention in Camden Town is called the Hawley Arms. Seeing it's where everything started, it felt right we should hold the book launch there. So we did. I hadn't been back since the midnineties—when Blur, and Oasis, and Britpop were all at the top of their game—and the place had certainly changed. But some things clearly *hadn't*. Later that evening, I had a dinner party to go to—and for one reason or another, never quite made it. You'd have thought I'd have learned my lesson.

In those days I wasn't married. Now I am. Thank God is all I can say. My wife, Elaine, has been the epitome of reason, the zenith of understanding, throughout this entire project—and her talent for cutting through crap (most of it mine) has proven invaluable. Just prior to publication, I decided to put my cards on the table. Elaine darling, I asked her (she could tell I was after something), Will you support me through my sex, drugs, and rock 'n' roll phase? I already have, she replied. Damn. I mean, thanks Elaine. And I love you.

In writing this book, it seems as if I've had more agents than MI5. Peter Tallack, Patrick Walsh, Clare Conville, Jake Smith Bosanquet, and Christy Fletcher have all performed heroically in keeping the show on the road—and when I managed to exhaust even *their* collective wisdom, Nick Kent kicked a few loose balls off the line (usually in glittering restaurants over the finest pinot noir). Don't know if anyone's ever put it quite like this before, Nick — but you're great between the posts.

I would also like to thank the following friends and colleagues for their advice and feedback during the course of writing this book. If I've missed anybody out, I'm afraid you're just going to have to face it — you weren't that important:

Dominic Abrams, Denis Alexander, Mike Anderson, Sue Armstrong, Phil Barnard, Michael Brooks, Peter Chadwick, Alex Christofi, Robert Cialdini, Max Coltheart, Keith Crosby, Jules Davidoff, Richard Dawkins, Roger Deeble, George Ellis, Ben Elton, Dan Fagin, Dan Gilbert, Andy Green, Cathy Grossman, Greg Heinimann, Paula Hertel, Rodney Holder, Emily Holmes, John Horgan, Stephen Joseph, Hubert Jurkiewicz, Herb Kelman, Deborah Kent, Linda Lantieri, Colin MacLeod, Bundy Mackintosh, Andrew Mathews, Ray Meddis, Ravi Mirchandani, Harry Newman, Pippa Newman, Richard Newman, Stephen Pinker, Martin Redfern, Russell Re-Manning, Gill Rhodes, V. S. Ramachandran, Jon Ronson, Jason Smith, Polly Stanton, John Timpane, Geoff Ward, Bob White, Mark Williams, and Konstantina Zougkou.

Special thanks also go to my editors at William Heinemann, Drummond Moir and Jason Arthur — two of the coolest, funniest, and nicest guys you could ever wish to work with — as well as to the equally wonderful Andrea Schulz and Tom Bouman at Houghton Mifflin Harcourt in the United States.

Extra special thanks goes to Sophie and Gemma (spelled with a J) Newman for hot flapjacks on cold winter Sunday afternoons.

And finally, this: On May 9, 1982, Hugh Jones toed the starting line of the London Marathon on Blackheath Common. Two hours, nine minutes, and twenty-four seconds later (as a fifteen-year-old, I remember watching that race on TV as if it were yesterday), he crossed the finish line some three minutes ahead of his nearest rival. A few years later, I got to know Hugh and his family in London; we used to run together in Regent's Park (he could never keep up with me), and then have dinner at his house in Camden. We became good friends, and have been ever since. That friendship, Hugh's spirit, and his wife Cheryl's Caribbean cooking, have stood me in pretty good stead over the years — not least during the writing of this book. Hugh, I just wanted to say thanks mate.

Illustration Credits

Figure 1.1 (p. 20) © Frank Greenaway/Dorling Kindersley Collection/Getty Images; Figure 1.3 (p. 23) © Keystone Features/Hulton Archive/Getty Images; Figure 1.4 (p. 24) © Mario Tama/Getty Images News/Getty Images; Figure 2.1 (p. 39) adapted from Sander, Frome, and Scheich, "FMRI Activations of Amygdala, Cingulate Cortex, and Auditory Cortex by Infant Laughing and Crying," *Human Brain Mapping* 28, no. 10 (2007): 1007–1022, reproduced by permission of John Wiley & Sons, Inc., and the authors; Figure 2.3 (p. 44) reprinted from Little and Hancock, "The Role of Masculinity and Distinctiveness in Judgments of Human Male Facial Attractiveness, *British Journal of Psychology* 93 (2002): 451–464, reproduced with permission from the *British Journal of Psychology* © the British Psychological Society; Figure 2.4a, left (p. 45) © Bob Grant/Hulton Archive/Getty Images; Figure 2.4a, right © Dave Allocca/Time & Life Pictures/Getty Images; Figure 2.4b (p. 45) © Kevin Winter/Getty Images Entertainment/Getty Images; Figure 2.5 (p. 46) adapted from K. Lorenz, "Part and Parcel in Animal and Human Societies: A Methodological Discussion, 1950," *Studies in Animal and Human Behavior, Volume 2* (London: Methuen, 1971); Figure 2.6 (p. 47) adapted from Pittenger and Shaw, "Aging Faces as Viscal-Elastic Events," *Journal of Experimental Psychology: Human Perception and Performance* 1, no. 4 (1975): 374–382, © 1975 by the American Psychological Association and adapted with permission; Figure 2.7 (p. 47) adapted from Pittenger and Shaw, "Perceptual Information for the Age Level of Faces as a Higher Order Invariant of Growth," *Journal of Experimental Psychology: Human Perception and Performance* 5, no. 3 (1979): 478–493, © 1979 by the American Psychological Association and adapted with permission; Figure 2.8 (p. 48) reprinted from Glocker, "Baby Schema Modulates the Brain Reward System in Nulliparous Women," *Proceedings of the National Academy of Sciences* 106, no. 22 (2009): 9115–9119, reproduced by permission of *Ethology* © Wiley-Blackwell; Figure 2.9 (pp. 50–51) courtesy of Dr. Chris Solomon and Dr. Stuart Gibson, University of Kent (original photograph of David Cameron © PA Photos); Figure 2.10 (pp. 51–52) reproduced by permission of Professor Allan Mazur, Syracuse University; Figure 2.11 (p. 57) © K. Dutton, adapted from Friesen and Kingstone, "The Eyes Have It! Reflexive Orienting Is Triggered by Nonpredictive Gaze," *Psychonomic Bulletin and Review* 5, no. 3 (1998): 490–495; Figure 2.12 (p. 58) reproduced by permission of Uta Frith, adapted from *Autism: Explaining the Enigma* (London: Blackwell, 1989); Figure 2.13 (p. 60) © Brigitte Sporrer/Cultura Collection/Getty Images; Figures 2.15

Index

AAB pattern, in humor and music, 39–40
abuse, domestic, 151–52
advertising
    anchoring/framing, 109–11
    emotional impacts, 21–22
    by newborns, 34
    and physical attractiveness, 50
    sexual stimuli, 22–23
    simple, effectiveness of, 164
    social proof in, 91–92
    Stephen Leacock on, 100
    use of suggestion, 100
affiliation, 67–68, 86–96, 161
afterimages study (Moscovici), 135–36
Ali, Muhammad, 1, 10, 165
Alicke, Mark, 103
Ambady, Nalini, 184
American Psychological Association, 59
amygdala. *See also* brain function
    dysfunctional, and risk-taking, 208,
        210, 225
    hot empathy and, 202–4
    response to infants' cries, 38–39
    response to surprise, incongruity, 177
anchoring, 108–10
Andersen, Susan, 43–44
Andrews, Brian, 200
Angelou, Maya, 129
animals
    communication skills, 15–17
    conflict diffusion behaviors, 26–27

persuasive powers, 30
antithesis, principle of (Darwin)
    in animals, 26–29, 40
    and the unexpected, 85, 192
antonomasia technique, 188
anxiety, anxiety disorder
    attentional training and, 246
    and cognitive bias modification, 245
    and the dot probe paradigm, 245–46
    effect of color pink, 66–67
    first and second order, 243–44
    thought process modification for,
        246–47
appeasement behaviors, 26–29, 40
Applewhite, Marshall, 142, 147
approach behaviors
    confidence and, 160–61, 184
    role of attitudes and beliefs, 77–79
    and social influence, 67, 68
    "virus of approach," 109–17
Aristotle, 191
Asch, Solomon
    conformity study, 88–89
    line study, 130, 133
    supplementary traits, 260
    word-choice study, 120
associative networks, 79
attention
    "attentional capture" study, 56
    attentional cueing, 56–58
    attentional training, 245–46

attention (*cont.*)
  cognitive distraction and, 70–77, 176
  evolutionary persuasion principles, 160
  manipulating, 69–70, 161
  and social influence, 66–68
attitudes and beliefs
  and anxiety disorder, 243–44
  availability heuristics, 82–83
  and brain states, 240–43
  changing, 7, 244–47, 272–73
  and cognitive bias modification,
    245–46
  and cognitive dissonance, 239–40
  and cognitive function, 77–79
  confidence and, 229–30
  and confirmation bias, 139–41
  delusions, 227–28
  epicenters, 275
  fixed vs. growth mind-sets, 231–33
  and interpretation of information,
    80–81
  performance and, 82–83, 230–31
  role in persuasion, 75
  unshakable, factors contributing to,
    229
attraction
  indicators of, 60–62
  of newborns, 41–43
attributional style, 148–50, 155–56
auctions, use of social proof in, 92–93
Audacity of Hope, The (Obama), 123
aural perception, 190
authority
  deference to, 167, 274
  and eye contact, 55
autism, 58–59
autosuggestion, 152–53
availability heuristic, 82–83, 85

babies. *See* infants, newborns
baboons, conflict diffusion rituals, 26–27
baby faces, persuasive power, 49–52. *See
  also* infants, newborns

Baines, David, 213
Baker-Miller pink, 67
Baron-Cohen, Simon, 205–7
Barrett, Keith
  church-donation hustle, 67–68
  on mental shortcuts, 80
  mind-reading skills, 195, 207
  persuasive powers, 65–66, 68
  "Three A's" of social influence, 67–68,
    91
Batson, Daniel, 169–71
Beachy Head Suicide Watch, 53–54
Bechara, Antoine, 209
bees, response to yellow, 20
beliefs. *See* attitudes and beliefs
Bem, Sandra, 43–44
bias, in-group, 87–91
bidding wars, 92–93
Bieber, David, 201
"big covers small" maxim, 173–74,
  179–80
birds, use of key stimuli, 16–17
Bizer, George, 107–8
blackmail, emotional, 112–14
blinking, 54–55
blueberries, mimicry of, 19
body language, 28–30
bomb-disposal experts, heart rates, 211
brain function/activity
  bias toward simplicity, 162
  during comparison procedure, 109–10
  complexity of, 80
  fixed vs. growth mind-sets and, 232
  following stroke, and rewiring, 248–49
  hot vs. cold empathy and, 202, 277
  imaging tools for, 201
  in low vs. high scorers on the PPI,
    204–5
  neural correlates to conformity, 89
  probabilistic inferences, 82
  relation to brain structure, 7
  response to belief, disbelief, and
    uncertainty, 240–43

response to cognitive bias
   modification, 248–49
response to cognitive distraction,
   70–77
response to newborns/infants, 34,
   37–39
response to color pink, 66–67
response to representativeness
   heuristic, 81
response to surprise, incongruity,
   177–79
when focusing attention, 69–70
brain images
   imaging tools, 201
   use of during court trials, 182–83
brainwashing, 133
Brandes, Eddo, 192
branding, emotional, 111
Breen, Nora, 228
Brochet, Frederic, 81
Brown, Gordon, 50–51, *50*
Brownlow, Sheila, 49–50
Buddhist monks, empathy in, 277
bullying, 152
Bundy, Ted, 207
Bush, George W., 126
Bush-Clinton debate conformity study,
   90–91
buzzwords, 124

Caine, Michael, 54–55
Cameron, David, 50–51
Cantor, Nancy, 141
Capgras delusion, 227
captor-captive bonding, 145–47
cartoon caricatures, 69
Castel, Alan, 182–83
cats, "solicitation purr," 15–16
CBM (cognitive bias modification),
   245–46, 249
central route to persuasion, 273
Chapman, Graham, 157–58
Chesterton, G. K., 55

Chiroxiphia pareola, 16
Churchill, Winston, 10, 29–30, 163–64
Cialdini, Robert
   compliance studies, 114, 134
   evolutionary persuasion principles, 167
   on limits of persuasion, 226–27
Cleese, John, 157–58
"climate change," reframing, 126
coercion, persuasion vs., 5
cognitive bias modification (CBM),
   245–46, 248–49
cognitive dissonance, 237–40
cognitive function. *See also* brain
   function
   attitudes and beliefs and, 77–79
   attention and distraction and, 70–77
   cognitive consistency, 115–17
   confidence and, 183–84
   heuristics and, 79–82, 85–86
cognitive suggestion. *See* framing and
   anchoring; suggestion
cold empathy, 202
Coltheart, Max, 227, 249
commitment, role in persuasion, 115–18,
   167
comparisons, anchoring and, 109–11
concessions and reciprocity, 114
conciliatory gestures, 147
conditioning, operant, 271
confidence
   approach behaviors, 184
   competence and, 183–84
   Jesus's, 29
   and the lack of emotion, 210–11
   role in persuasion, 10, 25–26, 181
   self-concept and, 229–31
   simplicity and, 165
   in successful con men, 68, 211–12
confidence (con) men. *See also* Barrett,
   Keith
   persuasive powers, 65–68, 211–12
   seduction techniques, 84–86
   use of salience, 186

confirmation bias, 139–41, 143–44
conflict diffusion techniques, 26–29
conformity
    factors promoting, 133, 154
    and group polarization, 130–33
    language promoting, 134–35
conformity studies, 88–91
consistency, cognitive, 115, 147, 167
Conspicuous Gallantry Cross, 219–20
control
    control delusions, 227
    dynamics of, 146–50
Conway, Luke, 163
Cooper, Ron, 159–60, 162, 173–74
cortical shock, 248
Cotard delusion, 228
courtrooms. See lawyers, trial
Cox, Trevor, 36
crayfish, conflict diffusion behaviors, 27
Crowley, James, 143–44
crying, by infants, as key stimulus, 35,
    37–39, 41
culpability, determinants of, 102–3
cults and cult leaders, 142, 151

Dalai Lama, 277
Dalton, Pam, 35–36
Darley, John, 81, 169–71
Darwin, Charles, 26–27. See also
    antithesis; natural selection
Davidson, Patricia, 213
Davidson, Richard, 277
Davis, Barbara, 176
Deal, David, 111
DeBruine, Lisa, 186–87
Dershowitz, Alan, 183
detail, attention to, 66, 68–69
Diagnostic and Statistical Manual of
    Mental Disorders (DSM IV),
    autism in, 59
DiCaprio, Leonardo, 45
diplomacy, 167
Dirty Dozen, The (movie), 219

disbelief, 233, 240–43
discomycete fungus, mimicry by, 19
disruption effect, 176
distraction, cognitive
    brain function and, 174–76
    distraction crimes, 74–75
    role in persuasion, 70–77, 234–35
"diversionary" persuasion, 76–77
dogs, conditioning experiments, 147–48
Doherty, Drayton, 12–13, 70
domestic abuse, 151–52
doomsday cult study, 237–38
dot probe paradigm, 245–46
dual process model of social influence,
    135–39
Dweck, Carol, 231–32

Eastwood, Brian, 248
eBay auctions, use of social proof, 92
Edwards, Donald, 27
EEG (electroencephalograph) recordings
    fixed vs. growth mind-sets, 232
    focused compassion, 277
    and response to the unexpected, 177
emotions
    lack of, benefits, 209–10
    psychopaths' ability to read, 206–7
    and risk aversion, 208
    role in persuasion, 111–14, 179,
        191–92
    similarity and salience, 186
empathy
    with animals, 15
    aural perception, 190
    body language, 29
    communicating, 17–18
    hot and cold, 202, 277
    in psychopaths, 200
    similarity and salience, 186–87
    and split-second persuasion, 10, 25–26,
        184–93, 234
    visual mimicry as, 18–19
    Zen masters, time perception, 189–90

Englich, Birte, 109
Entertaining Mr. Sloane (Orton), 212–13
"Environment: A Cleaner, Safer, Healthier
     America, The" (Luntz), 126
equity premium puzzle, 208
ethology, kin selection, 275
evolutionary development, 56–57
evolutionary persuasion principles
     (Cialdini), 167
exogenous attentional capture, 173
expectations. *See also* attitudes and
     beliefs
  perception and, 77–81
  performance and, 82–83, 230–31
  violating, role in persuasion, 40–41, 85
*Expression of the Emotions in Man and
     Animals, The* (Darwin), 26–27
external influences
  internal stimuli vs., 105
  impact on mind-sets, 233–34
  situational factors, 144–45
eye, human, physiology of, 59–60
eye contact
  among autistic children, 58–59
  among listeners vs. talkers, 55
  and black-white contrast, 62–63
  by newborns and infants, 35, 55–56
  role in persuasion, 54–55, 59–60

faces, baby, 49–52
facial electromyography (EMG), 192
facts, role in persuasion, 179, 272–73
Farroni, Teresa, 56
Far Side (Larson), 157
feminized males, appeal of, 44
Ferguson, Alex, 93
Festinger, Leon, 237–39
Field, Matt, 248
fixed vs. growth mind-sets, 231–33
fMRI (functional magnetic resonance
     imaging)
  effects of cognitive behavioral
     management, 249

  response to babies' faces, 48
  response to belief, disbelief and
     uncertainty, 240–43
  response to contradictions, 178, 191
  response to infants' cries, 38
  response to moral dilemmas, empathy,
     203–4, 207
  response to pleasure, 81
  use of in courtrooms, 182
focus, psychopaths' capacity for, 222
Foot, Philippa, 202
foot-in-the-door technique, 116–17
Ford, Henry, 229–30
Fox, Elaine, 248–49
framing and anchoring. *See also*
     suggestion
  language and word choice, 118–22
  reciprocity principle, 112–14
  role in persuasion, 107–9, 275
  use of in advertising, 111
Fraser, Scott, 116–17
Freedman, Jonathan, 116–17
Fregoli delusion, 228
frequency (sound) of infants' cries,
     37–38
Friedman, Ray, 229–30
Friesen, Chris, 56–57
frogs, use of key stimuli, 17
functional magnetic resonance imaging.
     See fMRI
functional psychopaths, 209
fundamental attribution error, 103–5

Gallagher, Liam and Noel (Oasis band),
     166–67
gang rape, 95
Gates, Bill, 210
Gates, Henry Louis, 143–44
Gauss, Carl Friedrich, 158
gender
  and performance, 230–31
  and physical appeal, 44
  and response to touch, 84

Gilbert, Dan, 233–34
Glocker, Melanie, 48
Golden Orb Weaver (spider), mimicry
     by, 20
Goldstein, Noah, 135
golfing, relationship between belief and
     performance, 230
Gordon, Heather, 204–5
Gospel of St. John, 27–29
Graduate Record Examination (GRE),
     expectations and performance, 82,
     230–31
Green, Andy, 151–52
Greene, Joshua, 203
Green treefrog, use of mimicry, 17–18
Gross, Paul, 81
Groth, Nicholas, 95
groups
   cohesion of, and synchronous activity,
     144
   and confirmation bias, 142–44
   and conformity, 153
   dependency on, polarizing impact,
     130–33, 146
   in-group bias, 87–91
   identification with, and male bonding,
     93–96
   need to belong to, 86–87
groupthink, 142–43
Guthrie, R. D., 63
gut instinct, 100

habits, mental, 77–79
Haigh, Tara, 201
Hairstreak Butterflies, mimicry by, 20
halo effect, 42
Hanaghan, Jonathan, 32
Hare, Robert, 199
Harris, Sam, 238–40
hearing, human, 37
Heath, Chip, 144
Heaven's Gate cult mass suicide, 142
helplessness, learned, 147–48

Hendy-Freegard, Robert, 211–12, 278
Herring Gull, response to key stimulus,
     23–24
heuristics
   and interpreting information, 79–81
   perceived self-interest and empathy
     as, 167
   representativeness heuristic, 80–83
   socioeconomic status (SES), 257–58
Hewstone, Miles, 144–45, 149
Hobbes, Thomas, 169
Hofstadter, Douglas, 168
honesty
   persuasive power of, 6
   University of Kent honesty study, 50
Hopkins, Anthony, 213
hot empathy, 202, 277
Hovland, Carl, 275
humor
   AAB pattern, 39–40
   impact on brain function, 177
   persuasive powers of, 174, 192
   and politeness, 218
Huron, David, 41
hypnosis, 70, 152–53

ideology and dogma, 130
implicatures, 217–19
impulse purchases, 111
impulsivity, 218–19
incongruity, surprise
   interference/disruption effect, 176
   reframing forced by, 177–79
   role in persuasion, 165, 171–80
   smiling as response to, 192
   and split-second persuasion, 10, 165
   use of, in captor-captive bonding, 147
   use of, by trial lawyers, 174
infants, newborns
   autistic, focus on mouth, 58
   black-white contrasts and, 62–64
   cry of, 37–38
   eye contact by, 55–56

persuasive powers, 32–35, 64
physical appeal, attractiveness, 44, 46–47
physiological response to, 38–39
inference, probabilistic, 82
influence. *See also* social influence
 continuum of, 229, 240–43, 250–52
 nonverbal, 14–20
information, presenting, 101–2
in-group bias, 87–91
"Innate Forms of Potential Experience, The" (Lorenz), 46
instinctive persuasion, 14–22
interference effect, 176
internal stimuli, 4, 105, 144–45
interrogators, successful, 235–36
investors, equity premium puzzle, 208
irises (eyes), human, 60
isolation, and group polarization, 133, 142
Issa, Fadi, 27

Janis, Irving, 142–43
Janiszewski, Chris, 109–10
Jesus Christ, 27–29, 241
Jones, Jim, 139, 144, 147, 228
Jones, Stanley E., 84
Jonestown, Guyana, mass suicide, 130, 142, 151
Jordan, Brent, 43
judgments of behavior, and fundamental attribution error, 103–5

Kampusch, Natascha, 145–47, 269
Keech, Marion, 237–38
Kennedy, John F., 179
key stimuli
 in conflict situations, 25–27
 enhanced, supernormal, 24
 role perception and social cognition, 96
 simplicity of, 163
 and stereotyping, 257–58

types and examples, 16–18, 23–24
 use of by newborns, 33–35, 64
Khan, Shaffiq, 84–86
kindchenschema (baby schema), 46–52
King, Martin Luther, 188
Kingstone, Alan, 56–57
kin selection, 275
Klucharev, Vasily, 89
Knowles, Eric, 176
Kringelbach, Morton, 34, 48

Lane, Keith, 52–54
Langer, Ellen, 73
language
 figurative, impact on empathy, 190–91
 role in persuasion, 30, 118–24
 role in promoting conformity/ compliance, 134–35
lap-dancer experiment, 43–44
Larson, Gary, 157–58
laughter, canned, 89
lawyers, trial
 legal arguments, 100, 250
 role of language and word choice, 122
 use of surprise, incongruity, 174
Leacock, Stephen Butler, 100
leadership, powerful/charismatic, 97, 133, 142, 151, 221
learned helplessness
 and attributional style, 148–50
 behavioral characteristics, 147–48
 bullying, domestic abuse and, 151–52
Lecter, Hannibal (fictional character), 199, 213, 217, 218. *See also* psychopaths
Lemann, Nicholas, 124–25
liking, role in persuasion/self-deception, 146, 167, 240
Lilienfeld, Scott, 200
limits of persuasion, 227–29
listening, eye contact during, 55
locus of control, internal vs. external, 149–50

Loftus, Elizabeth, 122
Lorenz, Konrad, 46
Lowenstein, George, 208
Luntz, Frank, 124–25, 126

Mackintosh, Bundy, 15, 16, 246–47
MacLeod, Colin, 243–46
magicians, misdirection by, 173–74
magnetoencephalography (MEG), 34, 201
majority influence, 89–91, 136
male attractiveness, predictors, 43–44
male bonding, 94–96
Manchester United football team, 93
Mancini, Marco, 25–26, 29
Mansfield, Michael, 100, 104–6, 250
marketing techniques
    appeal to emotions, 111–14
    foot-in-the-door approach, 116–17
    framing and anchoring, 109–12
    language and word choices, 119–20
    lowballing, 117–18
Marks & Spencer, 164
Mathews, Andrew, 246–47
McCabe, David, 182–83
McCain, John, 119
McComb, Karen, 15–16
McGlone, Matthew, 164
McGrath, Glenn, 192
MEG (magnetoencephalography), 34, 201
men
    attractiveness to women, 43–44
    male bonding, 94–96
    psychopathic disorder among, 198
Mencken, H. L., 225
mental set, 70–77
mental state, communicating, 59
Milgram, Stanley, 57–58, 117–18, 182, 274
military
    courage in battle, 220–21
    creating empathy for, 54
    detention centers, 67
    interrogators, successful, 77, 148,
        235–36

    use of confirmation bias, 144
Miller, Geoffrey, 43
mimicry, 17–20
mind control, 16, 147, 161. See also cults
    and cult leaders
mind reading, empathy and, 190
mind-sets, fixed vs. growth, 231–32
minimal group paradigm, 87
minority influence, 135–39
mirrored-self misidentification delusion,
    228
Mirror Man, 227–28, 237, 249
misdirection, 70–77, 173
MISS (Multidimensional Iowa
    Suggestibility Scale), 252–53
Monilinia vaccinii-corymbosi (fungus),
    mimicry by, 19
mood, altruistic behavior and, 192
moral dilemmas, 203
Morant, Greg, 181, 186, 213–14, 221
Moscovici, Serge, 135–39
Mozart, Wolfgang Amadeus, 40
Multidimensional Iowa Suggestibility
    Scale (MISS), 252–53
music, violation of expectation in, 40–41
Mussweiler, Thomas, 109
myopic loss aversion, 208

narratives, cohesive, persuasive power of,
    101, 104–5
natural selection
    and attraction to groups, 87
    and key stimuli, 18–19
    and limits of attention, 71, 106
    and minimal group paradigm, 87
    and survival of newborns, 34–35,
        37–38, 63
neuroeconomics, 208
neurosurgeons, 210
newborns. See infants, newborns
Newman, Richard, 94
9/11 terrorist attack, New York, 130, 134
nonverbal communication, 14–20, 55

numbers, rounded vs. precise, 109–10
nurse evaluations, 184

Oasis (rock band), refund checks from, 166–67
Obama, Barack, 119, 123, 188
odors, universal repellents, 35–36
Odyssey (Homer), 7–8
oil drilling, perceptions of, 125–26
operant conditioning, 271
optimists, attributional style, 149–50
oratory, persuasive, 191
order effect, 101–2
Orton, Joe, 212–13
Ottati, Victor, 190–91
Owl Butterflies, mimicry by, 19–20

paradoxical intention, 244
paralysis, rewiring the brain following, 248–49
parenting, 29–30. *See also* infants, newborns
passive avoidance learning tasks, 219
passive misdirection, 173
Pattinson, Robert, 45
Paul (friend), persuasive powers of, 222–24
PCL-R (Psychopathy Checklist–Revised), 199
peer pressure, 89–91, 94–95. *See also* groups
Peitho/Suadela (gods of persuasion), 14
perceived self-interest
    role in persuasion, 166–67, 169–71, 234
    and social proof, 167
    Wolf's Dilemma, 168–69
perception
    aural, 190
    impact of attitudes and beliefs, 77–81
    of key stimuli, 96
    and minority influence, 135–39
    self-perception, 145
    of trustworthiness, 50–52

performance
    and cognitive distraction, 174–76
    impact of self-concept/expectations on, 82, 230–31
peripheral route to persuasion, 273
Perrett, David, 43–44
persecution delusions, 227
persuasion. *See also* SPICE model of persuasion; "Three A's" of persuasion
    as continuum of influence, 10, 229, 240–43, 250–52
    exposure to, 4–5
    key stimuli, 15–18, 21, 23–24
    mistrust of, in popular imagination, 13–14
    split-second, overview, 9–11
pessimists, attributional style, 149–50
Pharisees, Jesus's encounter with, 27–29
phobias, first and second order, 243–44
Photuris fireflies, mimicry by, 21
physical attractiveness
    of infants, as key stimulus, 35, 46–49
    male, predictors, 43–44
    persuasive power of, 41–43
Piano Sonata in A Major (Mozart), 40
pink color, 66–67
Pinker, Steven, 217–19
pitch changes, 39
placebo effect, 141
plants, mimicry by, 19
Plassman, Hilke, 80–81
politeness, 218
political persuasion
    buzzwords, 124
    emotional oratory, 179
    emphasizing similarities, 188
    framing and anchoring, 107–9
    language and word choice, 121–23
    role of the cohesive narrative, 101
    simple rhetoric, effectiveness of, 163
    trustworthiness, 50–51
power differentials, 147

PPI (Psychopathy Personality Inventory), 200
presentations, effective, 130–33
Priklopil, Wolfgang, 146–47, 269–70
probabilistic inference, 82
Project Pigeon (Skinner), 271
psychopaths, psychopathic disorder
    characteristics, 197–99
    clinical, 55, 201
    cold vs. hot empathy in, 202–7
    continuum of behaviors, 199–200, 213, 220–21
    film portrayal, 212–13
    focusing capacities, 220–21
    persuasive powers, 200
    and self-interest, 219
Psychopathy Checklist–Revised (PCL-R), 199
Psychopathy Personality Inventory (PPI), 200
pupils, human eye, 60–62
purrs, variety of, 15–16
Pygmy Owls, mimicry by, 19

Rachman, Stanley, 211
racial prejudice, 123, 132, 143–44
Raine, Adrian, 219
Ramachandran, V. S., 41
Rasputin, Grigori, 188
Ratner, Gerald, 111–12
reading and cognitive function, 73–74
Reading the Mind in the Eyes test, 205–6
reciprocity principle, 112–14, 147, 167
reference delusions, 227
representativeness heuristic, 80–82, 85
resisting persuasion (unbelief), 234–37
restaurant no-shows, cancellation calls, 115–16
restaurant tips, influencing size of, 85, 115–17
Reynolds, Pat, 112–15
Rhea, Darryl K., 111
rhetoric, incisive, 7

rhyme, in poetry, 164
Right Touch: Understanding and Using the Language of Physical Contact, The (Jones), 84
risk-taking behaviors
    impact of group dynamics, 131–32, 142
    spectrum of, 132, 168, 208–9, 211, 221
rituals, 16, 117, 144
Roosevelt, Theodore, 125
Rosenthal, Robert, 184
Ross, Lee, 104–5
Rozin, Paul, 39

sales pitches, 7, 29
salience, and empathy, 17, 186
Sally Anne Task, 58–59
Sander, Kerstin, 38
Savill, Paul, 216–17
scarcity, role in persuasion, 167
Schauss, Alexander, 67
schema, associative networks, 46–52, 79
Seduction Community, 16
self-confidence. See confidence
self-interest. See perceived self-interest
self-perception, 59, 145
Seligman, Martin, 147–48, 150, 152
SERE (survival, evasion, resistance, escape) program, 148
7/7 terrorist attack, London, 134, 142
7Up can designs, 111
sex, as key stimulus, 22–24
sexual postures, submissive, 26–27
Sherif, Muzafer, 275
Shih, Margaret, 82, 230–31
Shiv, Baba, 208–9
shortcuts, persuasive powers of, 158–60
Silence of the Lambs (movie), 213, 217
"silent ringtone," 37
similarity
    brain's bias towards, 186–87
    empathy and, 186
    role in social influence, 188

simplicity
  brain's bias towards, 162
  confidence and, 165
  of key stimuli, 163
  smiling and, 192
Simpson, O. J., 183
Sinclair, Gordon, 115–16
Sirens, in the *Odyssey*, 7–8
Skinner, B. F., 271
Sloan, Vic, 66–67, 84
smiling, 34, 49, 167, 192
Smith, Frederick, 174
Snyder, Mark, 42, 141
social influence
  attention, approach and affiliation,
      67–68
  confirmation bias, 139–41
  "dual process" model, 135–39
  group polarization and, 130–33
  of newborns, 34–35
  politeness and, 217–18
  role of language and word choice,
      118–22
  similarity and, 188
  touch and, 84–85
  ubiquity of, 7
social phobias, treating, 240–43
social proof, 91–92, 167
Society, Evolution, and Revelation
      (Hanaghan), 32
socioeconomic status (SES) heuristics, 81,
      257–58
sound (acoustic stimuli)
  infants' cries, 37–38
  as key stimuli, 16–17
  "Mosquito" teenager repellent, 36–37
Spencer, Liam, 213–15
SPICE model of persuasion
  combined with focus, 222
  combining elements of, 161–62, 214–15
  Confidence, 27–29, 180–84
  Empathy, 184–93
  eye contact, 54–55

Incongruity, 171–80
  and neurological responses, 242
  overview, 10, 161–62
  and paradoxical intention, 244
  Perceived self-interest, 165–71
  Simplicity, 162–65
Spinoza, Benedict de, 233, 242
Stapleton, Howard, 36
statistics, communicating confidence
      using, 183
status inferences, 85
stereotypes, impact on performance, 82,
      230–31
stimuli, mental, 76–79. *See also* key
      stimuli
Stockholm syndrome, 146–47, 269
Stone, Jeff, 82, 229–31
Strack, Fritz, 109
Strohmetz, David, 85
stroke victims, treating, 248
Stroop Task, 176
Stuff of Thought, The (Pinker), 217
suggestion. *See also* framing and
      anchoring
  autosuggestion and learned
      helplessness, 152–53
  role in persuasion, 100, 107, 127
"suicide bumming," 96
surprise. *See* incongruity, surprise
*Sweet Anticipation: Music and the
      Psychology of Expectation* (Huron),
      41
Szot, Colleen, 91–92

Tajfel, Henry, 87
Tanweer, Shehzad, 132–33, 139
Taub, Edward, 248
teacher evaluations, 184
Thatcher, Margaret, 179
Theory of Mind (ToM), 58–59, 265
therapy, as form of persuasion, 244–45
"thin slicing" experiments, 184
Thomson, Judith Jarvis, 203

"Three A's" of persuasion (Barrett),
        160–61
Tofighbakhsh, Jessica, 164
torture, and learned helplessness, 148
touch (kinetic stimuli), 16, 84
trustworthiness, role in persuasion,
        49–51. See also confidence
Tybur, Joshua, 43

ultimatum game, 275
unbelief (resisting persuasion), 234–37
uncertainty, neurological response to,
        240–43
University of Kent honesty study, 50
U.S. Government Standard Bathroom
        Malodor, 36
Uy, Dan, 109–10

Vanders, Vance, 12–13
verbal communication, 7, 55, 118–24
Viagra, role in war on terror, 191
violence, threat of vs. actual, 236
visual stimuli, 4, 16
von Drehle, David, 123
Vonnegut, Kurt, 71

Westen, Drew, 178–79, 191
West Wing, The (television show), 119–20

Whitlam, Gough, 169
Williams, Pierre, 216–17
Willis, Bruce, 45
Wilson, Robert Anton, 181
Wiltermuth, Scott, 144
wine "quality" studies, 81
Wisdom of Forgiveness, The (Dalai Lama),
        277
Wiseman, Richard, 48–49
witchdoctors, 12–14
Wogan, Terry, 171
Wolf's Dilemma, 168–69
women
        eyebrow makeup, 52
        psychopathic disorder among, 198
        response to infants' cries, 38
words, persuasive powers of, 7, 118–24
World Cup, 2006, 145

young male syndrome, 94–95
Youth Offending Team, Cambridge, 94

Zarnoth, Paul, 183–84
Zebrowitz, Leslie, 49–50
Zen masters, 189–90